DATA MINING

Technologies, Techniques, Tools, and Trends

Bhavani Thuraisingham, Ph.D.

CRC Press
Boca Raton London New York Washington, D.C.

Library of Congress Cataloging-in-Publication Data

Thuraisingham, Bhavani M.
Data mining : technologies, techniques, tools, and trends / Bhavani Thuraisingham.
p. cm.
Includes bibliographical references and index.
ISBN 0-8493-1815-7
1. Database management. 2. Data mining. I. Title.
QA76.9.D3T456 1998
006.3—dc21 98-38523
CIP

Direct all inquiries to CRC Press LLC, 2000 Corporate Blvd., N.W., Boca Raton, Florida 33431.

International Standard Book Number 0-8493-1815-7
Library of Congress Card Number 98-38523
Printed in the United States of America 1 2 3 4 5 6 7 8 9 0
Printed on acid-free paper

PREFACE

Recent developments in information systems technologies have resulted in computerizing many applications in various business areas. Data has become a critical resource in many organizations, and therefore, efficient access to data, sharing the data, extracting information from the data, and making use of the information has become an urgent need. As a result, there have been many efforts on not only integrating the various data sources scattered across several sites, but extracting information from these databases in the form of patterns and trends has also become important. These data sources may be databases managed by database management systems, or they could be data warehoused in a repository from multiple data sources. To provide the interoperability as well as warehousing between the multiple data sources and systems, and to extract information from the databases and warehouses various tools are being developed.

The focus of this book is on extracting useful information, patterns, and trends from the various data sources managed possibly by database management systems. That is, this book will be devoted to the emerging technology called data mining. In general, data mining is the process of extracting information and patterns often previously unknown from large quantities of data using various techniques from areas such as machine learning, pattern recognition, and statistics. Data could be in files, relational databases, or other types of databases such as multimedia databases. Data may be structured or unstructured.

Quite a few good texts are now emerging in data mining. However, many of these texts are focusing on data mining techniques. These are the algorithms employed in data mining. However, one needs some technical background to understand these techniques. The purpose of this book is to discuss complex ideas in a way that can be understood by someone who wants a background in data mining. Technical managers as well as those interested in technology will benefit from this book. We focus on a data-centric view of data mining. That is, while statistical reasoning and machine learning techniques are important to data mining, if the data is not organized well, the results obtained will not be meaningful. Therefore, we repeatedly stress the importance of having good data for data mining. Toward this end, we discuss various data management technologies critical to data mining and show their relationship to mining.

This book is divided into three parts. Part I describes technologies for data mining. Without the underlying technologies such as database systems, warehousing, and machine learning, one cannot develop data mining as a technology area. Since data is critical to data mining, we spend considerable effort on discussing database systems technology as well as data warehousing. Note that much of the information on these data management and data warehousing technologies has been obtained from our previous book *Data Management Systems Evolution and Interoperation*. We also discuss the relationship of these data management and data warehousing technologies to data mining. Then we provide an overview of other supporting technologies such as machine learning, visualization, decision support, statistics, and parallel processing. Finally, we provide an overview of architectural support for data mining. In other words, while architectural support is a different kind of supporting technology from database systems or machine learning, various developments in architectures such as client-server architectures and distributed object management systems will play a role in data mining in the future.

Part II describes techniques and tools for data mining. It starts with a discussion of the various steps to data mining such as getting the data ready, carrying out the mining, pruning the results, and evaluating the outcomes. Then a discussion of the outcomes expected of data mining, approaches to mining, as well as an overview of various data mining techniques are provided. We also discuss at some length a particular data mining technique of interest to us and that is based on logic programming. Data mining is more or less an art at present. We believe that approaches such as those based on logic programming could provide some foundations to mining and make it a scientific discipline. Finally, a discussion of the various data mining tools are given. Note that the information on commercial products could soon be outdated due to the rapid developments in technology. Therefore, we urge the reader to keep up with the literature and contact the various vendors for up-to-date information.

Part III describes emerging trends in data mining. These include mining distributed, heterogeneous, and legacy data sources, mining multimedia data such as text, images, and video, mining data on the world wide web, metadata aspects of mining, and the privacy issues of data mining. While much of the data is still in structured databases, the amount of unstructured data is increasing. This data has to be mined. Furthermore, the databases could also be distributed. The explosion of the information on the world wide web will necessitate mining this information to make it more manageable for the user. Metadata plays a major role in mining. Finally, while data mining is a useful technology, the tools could be used by various people to get unauthorized information. Therefore, security and privacy is important.

Although our previous book *Data Management Systems Evolution and Interoperation* would serve as an excellent source of reference to this book, this book is fairly self contained. We have provided a reasonably comprehensive overview of the various background material necessary to understand mining. Furthermore, for mining a particular type of data such as text or images, we have tried to provide an overview of multimedia databases and then discuss the issues of mining such data. We have two appendices that provide some information on technologies such as data management and artificial intelligence, which are important to data mining.

We have tried to obtain information on products that are current. However, as mentioned earlier, vendors and researchers are continuously updating their systems and therefore, the information valid today may not be accurate tomorrow. We urge the reader to contact the vendors and get up-to-date information. Note that many of the products are trademarks of various corporations. If we know or have heard of such trademarks we have used all capital italic letters for the product when we first introduce it in this book. Again due to the rapidly changing nature of the computer industry, we encourage the reader to contact the vendors to obtain up-to-date information on trademarks and ownership of the various products.

We have tried our best to obtain references from books, journals, magazines, and conference and workshop proceedings, and avoided URLs to web pages as references. Although URLs contain excellent reference material, some of them may not be available even when this book goes into print. Therefore, we also encourage the reader to check the web from time-to-time for current information on data mining research, development, prototypes, and products. Three of the notable conference series in data mining are the Knowledge Discovery in Databases conference usually held in North America, Pacific Asia Knowledge Discovery in Databases conference usually held in Asia or Australia, and the Principles of Data Mining conference usually held in Europe.

We would like to stress to managers and project leaders not to rush into data mining. Data mining is not the answer to all problems. It is expensive and has to be thought out clearly. In reality, the mining portion is only a small part of the entire process. Is there a need for mining? Do you have the right data in the right form? Do you have the right tools? More importantly, do you have the people to do the work?

Do you have sufficient funds allocated to the project? All these questions have to be answered before you embark on a data mining project. Otherwise you can be extremely disappointed with the results.

We repeatedly use the terms data, data management, database systems and database management systems in this book. We elaborate on these terms in one of the appendices. We define data management systems to be systems that manage the data, extract meaningful information from the data, and make use of the information extracted. Therefore, data management systems include database systems, data warehouses, and data mining systems. Data could be structured data such as those found in relational databases, or it could be unstructured such as text, voice, imagery, and video. There have been numerous discussions in the past to distinguish between data, information, and knowledge. We do not attempt to clarify these terms. For our purposes, data could be just bits and bytes or it could convey some meaningful information to the user. We will, however, distinguish between database systems and database management systems. A database management system is that component which manages the database containing persistent data. A database system consists of both the database and the database management system.

This book provides a fairly comprehensive overview of data mining technologies, techniques, tools, and trends. It is written for technical managers and executives as well as for technologists interested in learning about data mining. I have been approached by various people to explain to them what data mining is all about. Therefore, I decided to write this book so that the complicated ideas can be expressed in a simplified manner and yet provide much of the information needed. This was also the reason for writing my previous book *Data Management Systems Evolution and Interoperation*. It should be noted that like many areas in data management, unless someone has practical experience in carrying out experiments and working with the various data mining tools, it is difficult to get an appreciation of what is out there and how to go about mining. Furthermore, unless one has practical experience, it will also be difficult to determine how well these tools operate. Therefore, we encourage the reader not only to read the information in this book and take advantage of the references mentioned here, but we also urge the reader, especially those who are interested in carrying out data mining, to work with the tools out there. Only then can the user have a good idea as to which tools are good for what outcomes and applications.

Data mining is still a relatively new technology. Although data management, statistical reasoning, and machine learning have been around for a while, it is only recently that these technologies are being integrated to mine data effectively. Therefore, as the various technologies and integration of these technologies mature, we can expect to see progress in data mining. That is, not only can we expect to mine various types of databases such as multimedia databases and web databases, we can also expect to extract useful information, patterns, and trends from these databases. This means that although we have tried to give a fairly comprehensive view of data mining covering many aspects, we can expect rapid developments with respect to many of the ideas, concepts, and techniques discussed in this book. We urge the reader to keep up with all the developments in this emerging and useful technology area. It should be noted that this book is intended to provide the background information as well as some of the key points and trends in data mining.

The views and conclusions expressed in this book are those of the author and do not reflect the views, policies, or procedures of the author's institution or sponsors. I thank my husband Thevendra for his encouragement and support. I thank my son Breman for our lively and intellectual discussions. I also thank my management for providing an environment where it is exciting and challenging to work, my professors and teachers for having given me the foundations upon which to build my skills, my sponsors and colleagues, all others who have supported my education and my work, and especially those who have reviewed various portions of this book.

Bhavani Thuraisingham, Ph.D.
Bedford, Massachusetts

About the Author

Bhavani Thuraisingham, Ph.D., recipient of IEEE Computer Society's prestigious 1997 Technical Achievement Award for her outstanding and innovative work in secure data management, is a senior principal engineer with the MITRE Corporation, Bedford, Massachusetts and heads the Data Management and Object Technology Department in the Information Technology Division. Previously, she held positions as lead engineer, principal engineer, and head of MITRE's research in Evolvable Interoperable Information Systems as well as Data Management, and co-director of MITRE's Database Specialty Group. Her current work focuses on data mining/knowledge discovery as it relates to text databases and database security, real-time multimedia database management, distributed object management, warehousing, and web databases.

Prior to joining MITRE in January 1989, Dr. Thuraisingham was a Principal Research Scientist with Honeywell Inc. conducting research, development, and technology transfer activities, and before that, was a Senior Programmer/Analyst with Control Data Corporation working on the design and development of the CDCNET product. She was also an adjunct professor of computer science and a member of the graduate faculty at the University of Minnesota. Dr. Thuraisingham earned an M.Sc. from the University of Bristol and Ph.D. from the University of Wales, Swansea, both in the United Kingdom. She is a senior member of the IEEE as well as a member of the ACM, IEEE Computer Society, the British Computer Society, and AFCEA. She has completed a management development program and is currently working toward a certification in Java programming.

Dr. Thuraisingham has published over three hundred technical papers and reports, including over forty journal articles, and is the holder of three U.S. patents for MITRE on database inference control. She also serves on the editorial boards of various journals, including IEEE Transactions on Knowledge and Data Engineering and the Journal of Computer Security. She gives tutorials in data management, including data mining and warehousing, object databases, distributed/heterogeneous databases, and Internet databases and currently teaches courses at both the MITRE Institute and the AFCEA Educational Foundation. She has chaired several conferences and workshops including IFIP's 1992 Database Security Conference, ACM's 1994 Multimedia Database Systems Workshop, IEEE's 1996 Metadata Conference, and IEEE's 1998 COMPSAC Conference. She is a member of OMG's real-time special interest group, and has served on numerous panels in data management and mining. She has edited several books as well as special journal issues in data management and object technology, and was the guest editor of the Data Management Handbook series by CRC's Auerbach Publications for 1996 and 1997. She is the author of the book *Data Management Sysems Evolution and Interoperation* by CRC Press.

Dr. Thuraisingham gives invited presentations at conferences, including the keynote address at the Second Pacific Asia Data Mining Conference as well as featured addresses at Object World East and West '96 and '97, ACM SAC Conference '97, IFIP Database Security Conference '96, Data Warehousing and Year 2000 Conference '96 and '97, and the IEEE Engineering Solutions Conference '96. She has also delivered the featured addresses at AFCEA's Federal Database Colloquiums from 1994 through 1998. She was instrumental in initiating AFCEA's Federal Data Mining Symposium series and co-chaired the first symposium in December 1997. Her presentations are worldwide, including in the United States, Canada, United Kingdom, France, Germany, Italy, Spain, Switzerland, Austria, Belgium, Sweden, Finland, The Netherlands, Greece, Ireland, India, Hong Kong, Taiwan, Japan, Singapore, New Zealand, and Australia. She also gives seminars and lectures at various universities around the world including at the University of Cambridge in England.

To My Dearest Husband Thevendra

Words cannot express the positive impact you have had on my life, education, and career;

and

To My Dearest Son Breman

Thank you for being a wonderful person and providing me with the motivation to continue my work.

TABLE OF CONTENTS

[1 2]

CHAPTER 1

INTRODUCTION

1.1 WHAT IS DATA MINING?

Data mining is the process of posing various queries and extracting useful information, patterns, and trends often previously unknown from large quantities of data possibly stored in databases. Essentially, for many organizations, the goals of data mining include improving marketing capabilities, detecting abnormal patterns, and predicting the future based on past experiences and current trends. There is clearly a need for this technology. There are large amounts of current and historical data being stored. Therefore, as databases become larger, it becomes increasingly difficult to support decision making. In addition, the data could be from multiple sources and multiple domains. There is a clear need to analyze the data to support planning and other functions of an enterprise.

Various terms have been used to refer to data mining as shown in Figure 1-1. These include knowledge/data/information discovery and knowledge/data/information extraction. Note that some define data mining to be the process of extracting previously unknown information while knowledge discovery is defined as the process of making sense out of the extracted information. In this book we do not differentiate between data mining and knowledge discovery. It is difficult to determine whether a particular technique is a data mining technique. For example, some argue that statistical analysis techniques are data mining techniques. Others argue they are not and that data mining techniques should uncover relationships that are not straightforward. For example, with data mining, a medical supplies company could increase sales by targeting certain physicians in its advertising who are likely to buy the products, or a credit bureau may limit its losses by selecting candidates who are not likely to default on their payments. Such real-world experiences have been reported in various papers (see, for example, [GRUP98]). In addition, data mining could also be used to detect abnormal behavior. For example, an intelligence agency could determine abnormal behavior of its employees using this technology.

Some of the data mining techniques include those based on rough sets, inductive logic programming, machine learning, and neural networks, among others. The data mining problems include classification (finding rules to partition data into groups), association (finding rules to make associations between data), and sequencing (finding rules to order data). Essentially one arrives at some hypothesis, which is the information extracted, from examples and patterns observed. These patterns are observed from posing a series of queries; each query may depend on the response obtained to the previous queries posed. There have been several developments in data mining. These include tools by corporations such as Lockheed Martin, Inc. (see, for example, [SIMO95]).

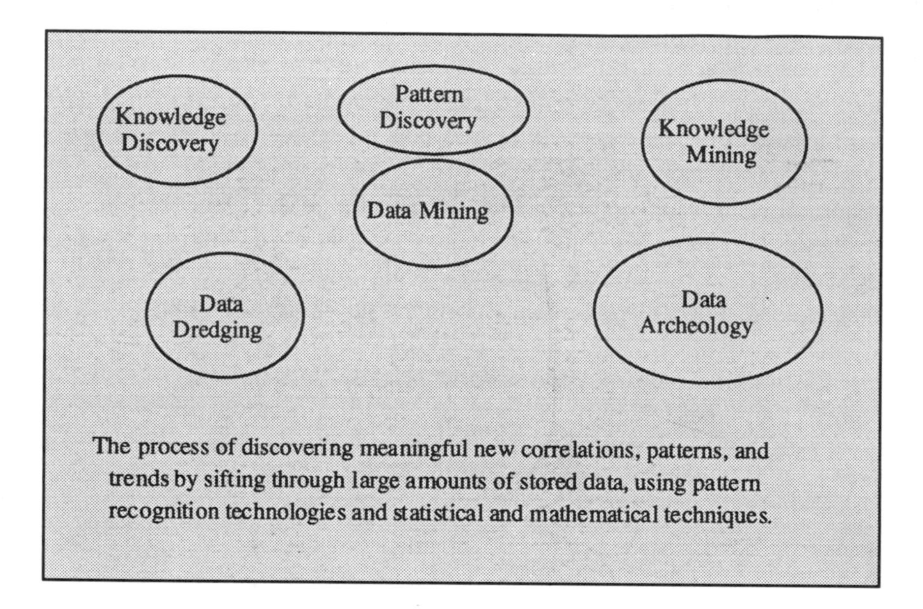

Figure 1-1. Different Definitions of Data Mining

This chapter is organized as follows. Technologies that contribute to data mining are discussed in Section 1.2. Essential concepts in data mining including techniques are discussed in Section 1.3. Trends in data mining is the subject of Section 1.4. Note that the contents of this book is an elaboration of each of the three Sections: 1.2, 1.3, and 1.4. In Section 1.5 we describe the organization of this book. Some additional discussions of the contents are given in Section 1.6 to give the reader better guidance. Section 1.7 gives suggestions for further reading.

1.2 DATA MINING TECHNOLOGIES

Data mining is an integration of multiple technologies as illustrated in Figure 1-2. These include data management such as database management, data warehousing, statistics, machine learning, decision support, and others such as visualization and parallel computing.[1] We briefly discuss the role of each of these technologies. It should however be noted that while many of these technologies such as statistical packages and machine learning algorithms have existed for many decades, the ability to manage the data and organize the data has played a major role in making data mining a reality.

Data mining research is being carried out in various disciplines. Database management researchers are taking advantage of the work on deductive and intelligent query processing for data mining. One of the areas of interest is to extend query processing techniques to facilitate data mining. Data warehousing is also another key data management technology for integrating the various data sources and organizing the data so that it can be effectively mined.

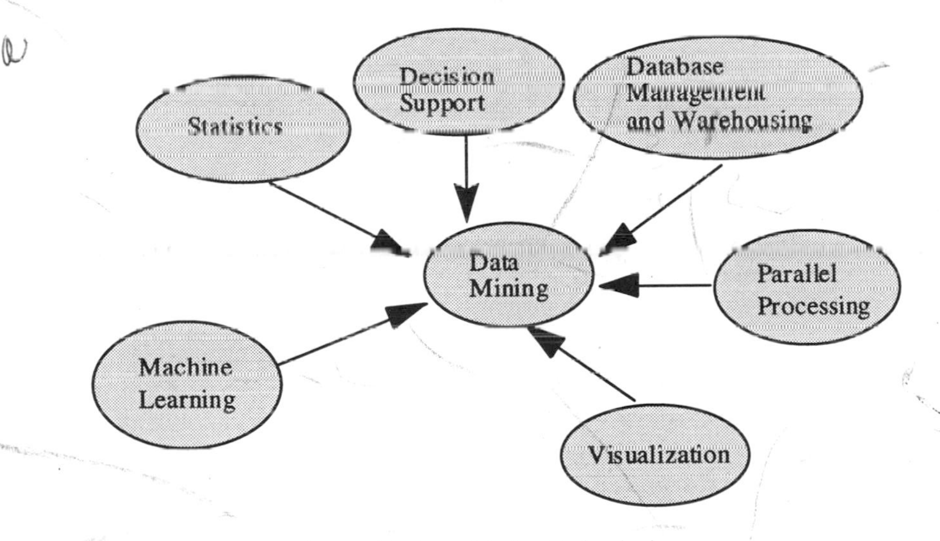

Figure 1-2. Data Mining Technologies

Researchers in statistical analysis are integrating their techniques with machine learning techniques to develop more sophisticated statistical techniques for data mining. Various statistical analysis packages are now being marketed as data mining tools. There is some dispute over this. Nevertheless, statistics is a major area contributing to data mining.

[1] We have distinguished between data management and database management and also between data, information, and knowledge. Our definitions are given in Appendix A as well as in [THUR97].

Machine learning has been around for a while. The idea here is for the machine to learn various rules from the patterns observed and then apply these rules to solve the problems. While the principles used in machine learning and data mining are similar, with data mining one usually considers large quantities of data to mine. Therefore, integration of database management and machine learning techniques are needed for data mining.

Researchers from the computer visualization field are approaching data mining from another perspective. One of their areas of focus is to use visualization techniques to aid the data mining process. In other words, interactive data mining is a goal of the visualization community.

Decision support systems are a collection of tools and processes to help managers make decisions and guide them in management. For example, tools for scheduling meetings, organizing events, spreadsheets view graph tools, and performance evaluation tools are examples of decision support systems. Decision support has theoretical underpinnings in decision theory.

Finally, researchers in the high performance computing area are also working on developing appropriate techniques so that the data mining algorithms are scalable. There is also interaction with the hardware researchers so that appropriate hardware can be developed for high performance data mining.

It should be noted that several other technologies are beginning to have an impact on data mining including collaboration, agents, and distributed object management. A discussion of all of these technologies is beyond the scope of this book. We have focused on some of the key technologies here. Furthermore, we emphasize that having good data is key to good mining.

1.3 CONCEPTS AND TECHNIQUES IN DATA MINING

There are a series of steps involved in data mining. These include getting the data organized for mining, determining the desired outcomes to mining, selecting tools for mining, carrying out the mining, pruning the results so that only the useful ones are considered further, taking actions from the mining, and evaluating the actions to determine benefits. These steps will be discussed in detail in this book. We briefly review some of the outcomes and techniques.

There are various types of data mining. By this we do not mean the actual techniques used to mine the data, but what the outcomes will be. Some of these outcomes are discussed in [AGRA93], and we will elaborate them in this book. They have also been referred to as data mining tasks. We describe a few here.

In one outcome of data mining, called "classification," records are grouped into some meaningful subclasses. For example, suppose an automobile sales company has some information that all the people in its list who live in City X own cars worth more than 20K. They can then assume that even those who are not on their list, but live in City X can afford to own cars costing more than 20K. This way the company classifies the people living in City X.

A second outcome of data mining is "sequence detection." That is, by observing patterns in the data, sequences are determined. Here is an example of sequence detection: after John goes to the bank, he generally goes to the grocery store.

A third outcome of data mining is "data dependency analysis." Here, potentially interesting dependencies, relationships, or associations between the data items are detected. For example, if John, James, and William have a meeting, then Robert will also be at that meeting. It appears it is this type of mining that is of much interest to many.

A fourth outcome of mining is "deviation analysis." For example, John went to the bank on Saturday, but he did not go to the grocery store after that. Instead he went to a football game. With this type, anomalous instances and discrepancies are found.

As mentioned earlier, various techniques are used to obtain the outcomes of data mining. These techniques could be based on rough sets, fuzzy logic, inductive logic programming, or neural networks, among others, or they could simply be some statistical technique. We discuss these techniques later in this book. Furthermore, different approaches have also been proposed to carry out data mining including top-down mining as well as bottom-up mining. Data mining outcomes, techniques and approaches are illustrated in Figure 1-3 and will be elaborated later in this book.

Numerous developments have been made in data mining over the past few years. Many of these focus on relational databases. That is, the data is stored in relational databases and mined to extract useful information and patterns. We have several research prototypes and commercial products. The research prototypes include those developed at IBM's (International Business Machines) Almaden Research Center and at Simon Fraser University. The prototypes and products employ various data mining techniques including neural networks, rule based reasoning, and statistical analysis. The various data mining tools in the form of prototypes and products will also be discussed in this book.

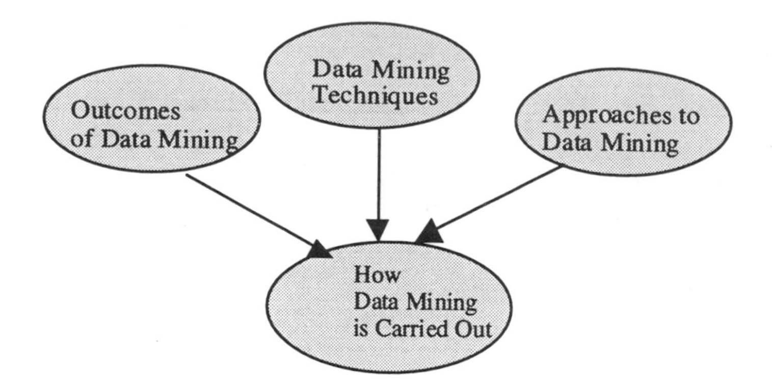

Figure 1-3. Aspects of Data Mining

1.4 DIRECTIONS AND TRENDS IN DATA MINING

While several developments have been made, there are also many challenges. For example, due to the large volumes of data, how can the algorithms determine which technique to select and what type of data mining to do? Furthermore, the data may be incomplete and/or inaccurate. At times there may be redundant information, and at times there may not be sufficient information. It is also desirable to have data mining tools that can switch to multiple techniques and support multiple outcomes. Some of the current trends in data mining include the following and are illustrated in Figure 1-4:

- Mining distributed, heterogeneous, and legacy databases
- Mining multimedia data
- Mining data on the world wide web
- Security and privacy issues in data mining
- Metadata aspects of mining

In many cases the databases are distributed and heterogeneous in nature. Furthermore, much of the data is in legacy databases. Mining techniques are needed to handle these distributed, heterogeneous, and legacy databases. Next, current data mining tools operate on structured data. However, there are still large quantities of data that are unstructured. Data in the multimedia databases are often semistructured or unstructured. Data mining tools have to be developed for multimedia databases. Next, the explosion of data and information on the world wide web necessitates the development of tools to manage and mine the data so that only useful information is extracted. Therefore, developing mining tools for the world wide web will be an important area. Privacy issues are becoming critical for data mining. Users now have sophisticated tools to make inferences and deduce information to which they are not authorized. Therefore, while data mining tools help solve many problems in the real world, they could also invade the privacy of individuals. Throughout our previous book [THUR97] we repeatedly stressed the importance of metadata for data management. Metadata also plays a key role in data mining.

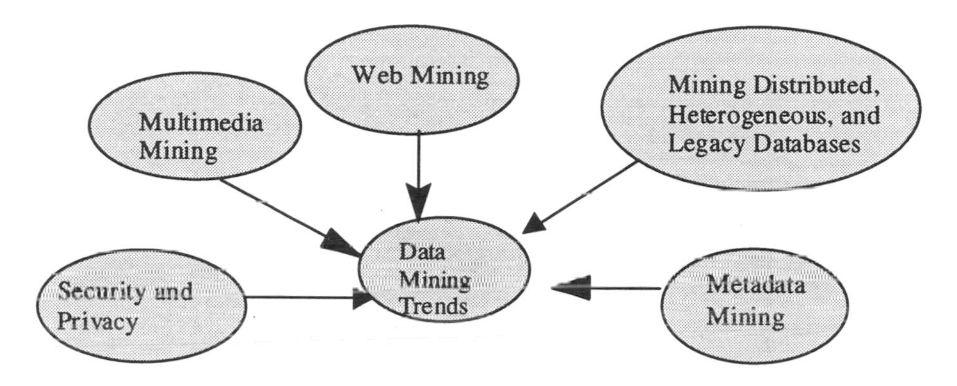

Figure 1-4. Data Mining Trends

In addition to the trends in the above areas, there are also several challenges. These include handling dynamic data, sparse data, incomplete and uncertain data, as well as determining which data mining algorithm to use and on what data to operate. In addition, mining multiple languages is also a challenge. Researchers are addressing these challenges.

1.5 ORGANIZATION OF THIS BOOK

This book covers the three essential topics in data mining in three parts: Technologies, Techniques and Tools, and Trends. To explain our ideas more clearly, we illustrate a data mining framework in Figure 1-5. This framework has three layers. Layer 1 is the Technologies layer. It describes the various technologies that contribute to data mining. These include data management, machine learning, statistics, visualization and parallel processing. In addition, architectural support for mining is also addressed. Layer 2 is the Techniques and Tools layer. This layer describes the various concepts, techniques, and data mining tools and makes use of the underlying technologies. For example, in the case of machine learning technology, one could utilize machine learning techniques such as neural networks for data mining. Layer 3 is the Trends layer, and this layer describes the recent trends in data mining such as mining distributed databases, mining multimedia data, mining the world wide web, metadata mining, and security and privacy issues.

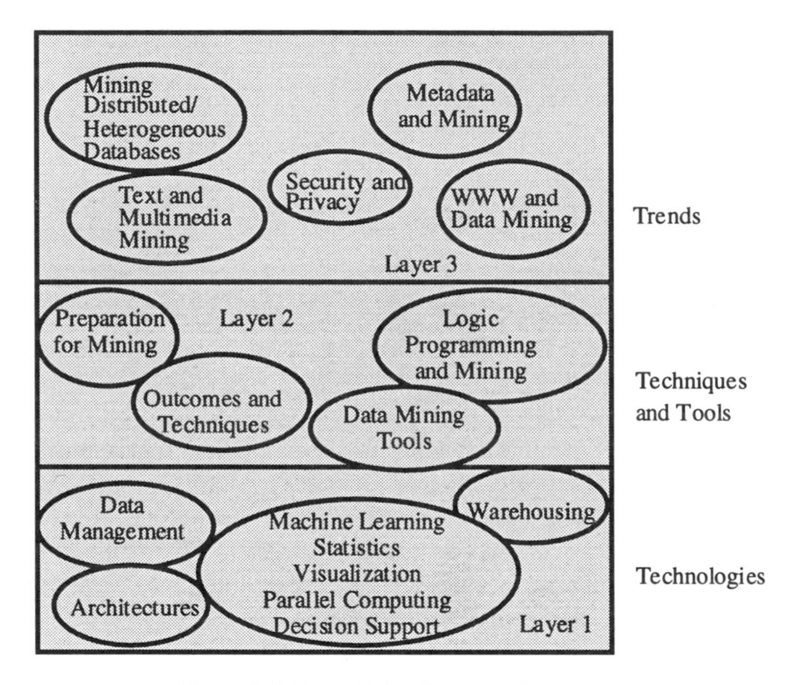

Figure 1-5. Data Mining Framework

Each layer is described in a part of this book. Part I, consisting of four chapters, describes the various data mining technologies. Chapter 2 describes database systems technology. This includes a discussion of data modeling, architecture, and functions. It should be noted that having good data is key to good mining. We believe it is the developments in database technologies that have mainly contributed to the existence of data mining as a technical area now. Chapter 3 describes a particular aspect of data management that is critical for data mining and that is data warehousing. To effectively mine the data, the data has to be formatted and organized. Warehousing is a way to accomplish this. We describe data warehousing concepts, developments, and relationships to mining. Since having good data is a critical part of mining, we have devoted two chapters to this topic. Chapter 4 discusses some other key technologies for mining. These include machine learning, visualization, decision support, and parallel computing. In addition, statistical reasoning techniques will also be described as they play a major role in data mining. Chapter 5 provides an overview of architectural support for data mining. It describes architectures for integrating with other technologies, functional architectures, and client-server-based architectures.

Part II, consisting of four chapters, describes concepts, techniques, and tools in data mining. Chapter 6 describes the steps involved in data mining. These include getting ready for mining, doing the mining, pruning the results, and evaluating the outcomes. Various examples where data mining could be utilized are also given so the reader can have a better appreciation for mining. Chapter 7 describes the essential points in data mining including the outcomes one could expect, the approaches used, and the techniques employed. Chapter 8 describes a particular data mining technique of interest to us that is based on logic programming. Currently we believe that data mining is very much an art. If data mining is ever to become a science we need to examine the foundations. Logic programming is one way to approach data mining as a scientific discipline. Finally, in Chapter 9 we discuss various data mining tools. These tools employ the concepts and techniques discussed in Chapter 7. The tools are in the form of prototypes as well as commercial products. It should be noted that many of the tools operate on structured databases such as relational databases. Furthermore, information on the products could soon become outdated. To get up-to-date information on the products, we encourage the reader to contact the vendors.

Part III, consisting of five chapters, describes trends in data mining. Chapter 10 describes mining distributed, heterogeneous, and legacy data sources. As discussed in [THUR97], many databases are getting distributed. Furthermore, the heterogeneous databases have to interoperate with each other. Finally, much of the data resides in legacy databases. The challenge is to mine such databases and still get some useful results. A brief discussion of collaborative data mining will also be given in this chapter. Chapter 11 describes mining multimedia databases such as text, images, and video. With large quantities of data now in nonrelational databases, mining unstructured databases, as well as multimedia databases, will be necessary in the future. Mining data on the world wide web is the subject of Chapter 12. We expect this area to develop rapidly over the next few years. While data mining technologies have been applied in a positive aspect, there are also some security implications. Users now have these data mining tools with which they can make correlations that are not desirable. Security and privacy issues are discussed in Chapter 13. The role of metadata in mining is becoming prominent, so we have devoted the entire Chapter 14 to discuss metadata aspects of data mining.

Figure 1-6 illustrates the chapters in which the components of the framework in Figure 1-5 are addressed in this book. We summarize the book and provide a discussion of challenges and directions in Chapter 15. Each of the chapters in Parts I, II, and III, in other words Chapters 2 through 14, starts with an overview of the chapter and ends with a summary of the chapter. Finally, we have two appendices that provide useful background information. As we will see, both data management and artificial intelligence technologies play a major role in data mining. In Appendix A, we provide an overview of the developments and trends in data management technology, and in Appendix B, we provide an overview of the developments and trends in artificial intelligence technology. We also provide a fairly comprehensive list of references in the section on references. We have obtained these references from various journals, conference and workshop proceedings, and magazines. In addition, each appendix also has its own set of references. We end this book with an Index.

1.6 ADDITIONAL DISCUSSION OF THE CONTENTS

One point to note is that the chapters in this book not only discuss data mining concepts, they also show how data mining could be applied to the various systems based on technologies of the data management framework we have discussed in our book, *Data Management Systems Evolution and Interoperation*. This framework is also a three-layer framework and is illustrated in Figure A-5 of Appendix A.[2] For example, Chapter 2 describes the relationships between mining and databases. Chapter 3 describes the relationships between mining and warehousing. Chapter 10 describes relationships between mining and distributed, heterogeneous, and legacy databases. Chapter 11 describes relationships between mining and multimedia data. Chapter 12 describes relationships between mining and the world wide web, which includes digital libraries and Internet databases. In essence, many of the technologies discussed in the framework of Figure A-5 given in the Appendix have been useful in the discussion of mining. These include database systems, distributed database systems, interoperability of heterogeneous database systems, migrating legacy databases,

[2] It should be noted that we are discussing two frameworks. One is a framework for data management that we proposed in [THUR97] and illustrated in Appendix A of this book. The other is a data mining framework that we illustrated in figure 1-5.

multimedia database systems, data warehousing, and digital libraries and Internet database management. In addition, some of the other features in data management such as metadata, security, and logic programming also play a role in various chapters of this book. For example, metadata and mining is the subject of Chapter 14. Security and Privacy issues is the subject of Chapter 13. Logic programming as a data mining technique is the subject of Chapter 8. Therefore, much of the discussions in this book have a strong orientation toward data and data management.

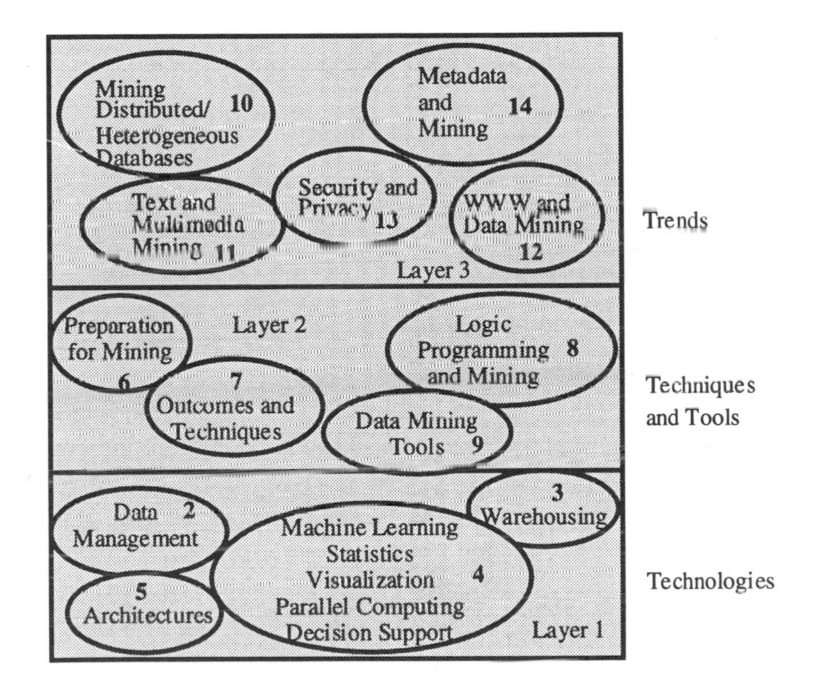

Figure 1-6. Components Addressed in this Book

While data is the main concern for us, we have not ignored some of the other essential technologies and features of data mining. For example, Chapter 4 discusses other data mining technologies such as machine learning and statistics. Chapter 5 discusses architectures for mining, Chapter 6 discusses steps to data mining. Chapter 8 describes various data mining concepts and techniques. Chapter 9 provides an overview of the data mining tools. We have tried to give a fairly balanced view of what is out there in data mining. Since artificial intelligence technology has also contributed much to data mining, we address this in Appendix B. In Figure 1-7 we illustrate how data management, machine learning, and statistical reasoning contribute to data mining. Data management provides data and the underlying middleware. Machine learning and statistical reasoning provide support to the techniques in data mining. Figure 1-8 illustrates how the components from the various layers can be put together to form a data mining

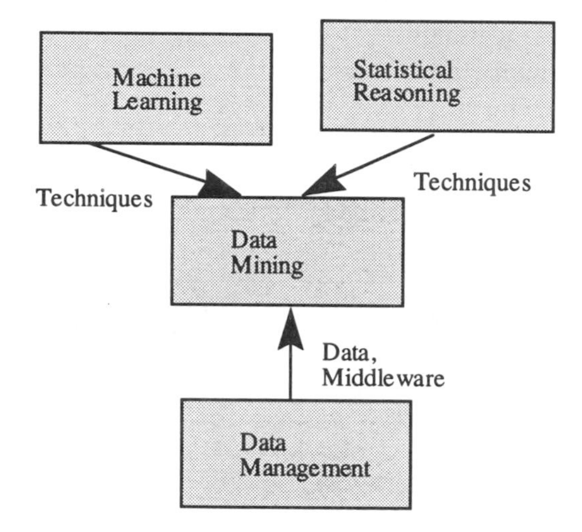

Figure 1-7. Contribution of Data Management, Machine Learning, and Statistics

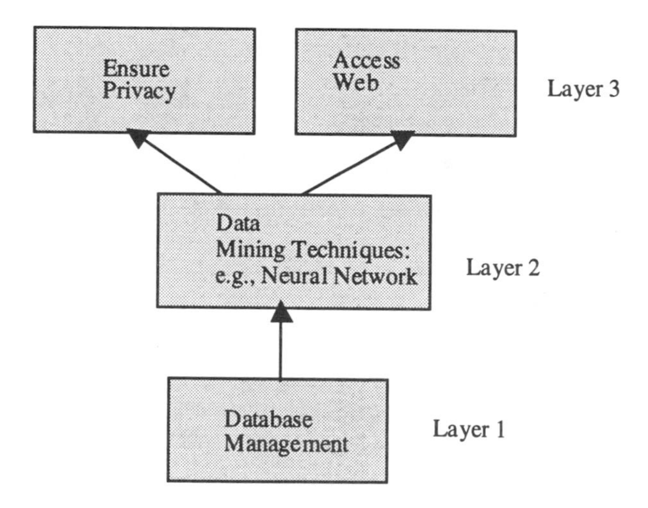

Figure 1-8. Layered Mining

system. For example, at Layer 1 we have database management, at Layer 2 we have various data mining techniques such as neural networks, and at Layer 3 we have web access and privacy issues.

Essentially, data mining can be applied to any type of database including relational and object databases, data warehouses, distributed and heterogeneous databases, multimedia databases, and web databases. This book is all about mining different types of databases. This is illustrated in Figure 1-9.

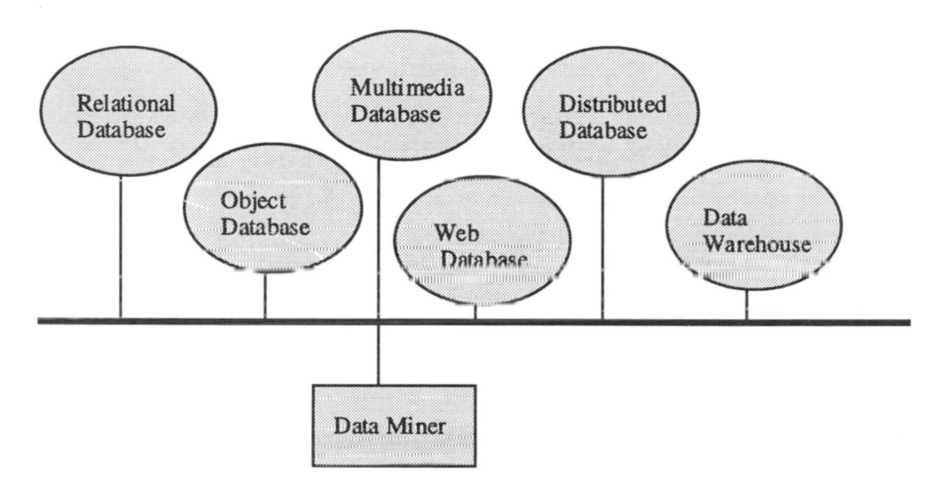

Figure 1-9. Mining Different Types of Databases

Data mining is very much an evolving field. Although research in machine learning, statistics, and database management has gone on for a long time, and quite a few of the data mining techniques have existed for decades, data mining emerged as a field in the late 1980s and early 1990s. This is partly due to the "knowledge discovery in databases" workshop series that started in the late 1980s and then evolved into the "knowledge discovery in databases" conference series. We illustrate the evolution of data mining in Figure 1-10. The question now is where does data mining go from here? We believe that as many of the issues and challenges described in this book are addressed, data mining will become a mature field. The ultimate goal, that has been around for a while, is not only for machines to mimic humans, but also to carry out tasks that would be difficult for humans to do.

1.7 HOW DO WE PROCEED?

This chapter has provided an introduction to data mining. We first discussed various technologies for data mining, and then we provided an overview of the concepts in data mining. These concepts include the outcomes of mining, the techniques employed, and the approaches used. The directions and trends, such as mining heterogeneous data sources, mining multimedia data, mining web data, metadata aspects, and privacy issues, were addressed next. Finally, we illustrated a framework for data mining and showed how we address the components of this framework in this book.

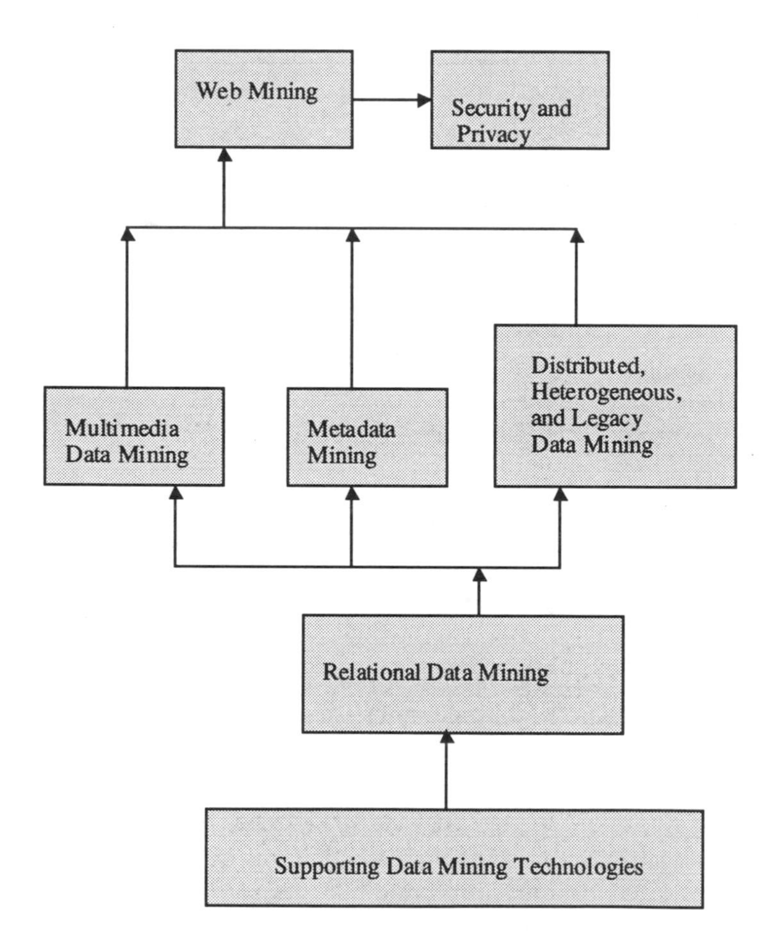

Figure 1-10. Evolution of Data Mining

Our framework is a three-layer framework and each layer is addressed in a part of this book. These are data mining technologies, data mining techniques and tools, and data mining trends. The next several chapters of this book will address the various technologies, techniques and tools, and trends in data mining. As it will be stressed throughout this book, having good data is key to good data mining.

This book provides the information for a reader to get familiar with data mining. Many important topics are covered so that the reader has some ideas as to what data mining is all about. For an in-depth understanding of the various topics covered in this book, we recommend the reader to the various references we have given. Numerous papers and articles have appeared in data mining and related areas. We reference many of these throughout this book. Some of the interesting discussions on data mining have appeared in [TKDE93], [TKDE96], [SIGM96], [KDD95], [KDD96], [KDD97], [KDD98], [PAKD97], and [PAKD98]. In addition, data mining papers have also appeared at various data management conferences (see, for example, [DKE98], [SIGM98], and [VLDB98]). Recently, a federal data mining symposium series has been established [AFCE97]. Note that many of these papers also take a rather data-centric approach to mining. There are, however, excellent articles emerging on the machine learning perspective. An example is the work of Mitchell [MITC97]. Furthermore, Berry and Linoff [BERR97] explain various types of data mining techniques clearly in their book. We expect data mining to be a multidisciplinary technology area with closer interactions and collaborations between the various communities such as data management, machine learning, and statistics to produce good data mining techniques and tools.

Part I

Data Mining Technologies

Introduction to Part I

Part I consists of four chapters describing the technologies that contribute to data mining. Chapter 2 discusses database management systems. These are systems that organize, structure, and manage the data. Having good data is key to getting good results from mining. Therefore, we focus quite a lot on database systems. Chapter 3 discusses data warehousing technology. A data warehouse integrates data from multiple data sources to aid in decision support functions. Having a good data warehouse makes it easier to mine the data. While database systems and data warehouses are two of the key technologies for data mining, some other technologies are also equally important. We discuss many of these other supporting technologies in Chapter 4. In particular, statistical reasoning, machine learning, visualization, parallel processing, and decision support technologies are discussed. In Chapter 5 we provide an overview of architectural support for data mining. In particular, architectures for integration, functional architectures, and client-server architectures are discussed.

The chapters in Part I provide the background material for data mining. They provide the foundations upon which the remaining chapters in this book are built. We have provided several references should the reader need more information on these topics. Much of the information on data management and data warehousing has been obtained from our previous book *Data Management Systems Evolution and Interoperation*.

CHAPTER 2

DATABASE SYSTEMS TECHNOLOGY

2.1 OVERVIEW

Database systems play a key role in data mining. Having good data is key to mining, and therefore, we give considerable attention to database systems in this book. It should be noted that we are taking quite a data-oriented perspective to mining.

Database systems technology has advanced a great deal during the past four decades from the legacy systems based on network and hierarchical models to relational and object-oriented database systems based on client-server architectures. This chapter provides an overview of the important developments in database systems relevant to the contents of this book. Much of the discussion in the remainder of this book builds on the information presented in this chapter.

As stated in Appendix A, we consider a database system to include both the database management system (DBMS) and the database (see also the discussion in [DATE90]). The DBMS component of the database system manages the database. The database contains persistent data. That is, the data is permanent even if the application programs go away.

The organization of this chapter is as follows. In Section 2.2 data models are described. In particular, relational, entity-relationship, object-oriented, object-relational, and logic-based data models are discussed. We focus on these models as they are discussed in later chapters with respect to mining. In Section 2.3 various types of architectures for database systems are described. These include an architecture for a centralized database system, schema architecture, as well as functional architecture.[3] Database design issues are discussed in Section 2.4. Database administration issues are discussed in Section 2.5. Database system functions are discussed in Section 2.6. These functions include query processing, transaction management, metadata management, storage management, maintaining integrity and security, and fault tolerance. Finally, in Section 2.7, we discuss data mining and its relationship to database systems. In particular, architectural issues, impact of data modeling, administration, design, and functions on

---

[3] We discuss distributed databases in the chapter on mining distributed and heterogeneous databases.

mining are discussed in some detail. The chapter is summarized in Section 2.8.

2.2 DATA MODELS

2.2.1 Overview

It is widely accepted among the data modeling community that the purpose of a data model is to capture the universe that it is representing as accurately, completely, and naturally as possible [TSCI81]. While philosophers have been interested in various types of universes for centuries, recently, it has interested the data model, logic, and database researchers a great deal. In particular, the work of Gallaire, Minker, Kowalski, Nicolas, Reiter, and Clark among others has described the differences between the actual and perceived universes, and they have developed approaches for modeling the perceived universe (see, for example, [GALL78], MINK88], [FROS86], and [BROD84]).

The actual universe has the truth about all of the entities in the universe. The perceived universe is the people's view of the universe. This view is usually determined by someone or a group of people in authority. One can regard the perceived universe to be an interpretation of the actual universe. Whether the perceived universe is a model of the actual universe depends on how accurately the perceived universe fits the actual universe. For data modeling purposes, it is the perceived view of the universe that is of interest. This is because the views of the users of the database must be correctly reflected. Figure 2-1 illustrates what has been discussed here.

Research in data modeling has concentrated mainly in two areas from which two types of data models were developed: traditional and semantic models. Traditional models include the network, hierarchical, and relational models (see, for example, [DATE90]). Semantic models include the entity-relationship [CHEN76] (also referred to as the ER model), functional [BUNE82], logic-based [ULLM88], and object-oriented models [BANE87]. Models such as the relational model are being extended to support complex objects. For example, relational and object-oriented models have been integrated to produce object-relational data models [ACM91]. Recently, another type of model has been included in the classification of data models. This is the hyper-semantic data model [TRUE89]. Such a model not only includes the constructs provided by semantic data models, but provides inference capabilities necessary to model knowledge-based applications. That is, hypersemantic data models integrate the constructs of semantic and logic-based models. Although semantic data models also provide powerful constructs (e.g., inheritance, generalization, aggregation, and composition), these are insufficient to model knowledge-based applications, which require inferencing capabilities.[4]

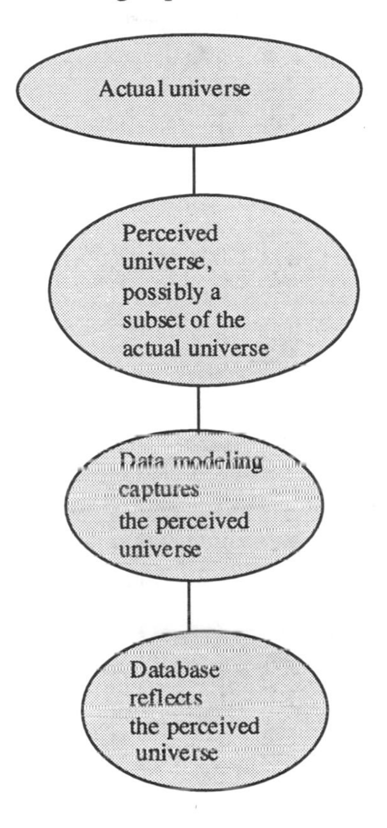

Figure 2-1. Data Modeling

In this section we discuss the essential points of relational, entity-relationship, object-oriented, object-relational, and logic-based data models. We have chosen these models because presently they are of

---

4 Note that many of the legacy systems which still dominate the business world are based on network and hierarchical models. However, many of the new initiatives are choosing the relational approach. Object-based approaches are being used for some applications. Logic-based approaches have not really taken off, but they are of much interest to the research community. Inductive logic programming has evolved from logic programming and logic-based approaches and is discussed in Part II of this book. In a functional model, the database is a collection of functions and query processing amounts to function execution. This is also an approach that has not taken off. Semantic models have evolved from ER as well as some of the artificial intelligence models.

much interest to the database community and have some impact on mining. For example, much of the data that is being mined is stored in relational databases. Entity relationship models are being used to design the database and are being examined for designing the warehouse schema. Object and object-relational approaches are becoming popular for multimedia databases and there is much interest to mine multimedia data. Finally, logic-based models may be useful for data mining based on inductive logic programming.

2.2.2 Relational Data Model

With the relational model [CODD70], the database is viewed as a collection of relations. Each relation has attributes and rows. For example, Figure 2-2 illustrates a database with two relations EMP and DEPT. EMP has four attributes: SS#, Ename, Salary, and D#. DEPT has three attributes: D#, Dname, and Mgr. EMP has three rows, also called tuples, and DEPT has two rows. Each row is uniquely identified by its primary key. For example, SS# could be the primary key for EMP and D# for DEPT. Another key feature of the relational model is that each element in the relation is an atomic value such as an integer or a string. That is, complex values such as lists are not supported.

EMP

SS#	Ename	Salary	D#
1	John	20K	10
2	Paul	30K	20
3	Mary	40K	20

DEPT

D#	Dname	Mgr
10	Math	Smith
20	Physics	Jones

Figure 2-2. Relational Database

Various operations are performed on relations. The SELECT operation selects a subset of rows satisfying certain conditions. For example, in the relation EMP, one may select tuples where the salary is more than 20K. The PROJECT operation projects the relation onto some attributes. For example, in the relation EMP one may project onto the attributes Ename and Salary. The JOIN operation joins two relations over some common attributes. A detailed discussion of these operations is given in [DATE90] and [ULLM88].

Various languages to manipulate the relations have been proposed. Notable among these languages is the ANSI Standard SQL (Structured Query Language). This language is used to access and manipulate data in relational databases [SQL3]. There is wide acceptance of this standard among database management system vendors and users. It supports schema definition, retrieval, data manipulation, schema manipulation, transaction management, integrity, and security. Other languages include the relational calculus first proposed in the Ingres project at the University of California at Berkeley [DATE90]. Another important concept in relational databases is the notion of a view. A view is essentially a virtual relation and is formed from the relations in the database. For further details we refer to [DATE90].

2.2.3 Entity-Relationship Data Model

One of the major drawbacks of the relational data model is its lack of support for capturing the semantics of an application. This resulted in the development of semantic data models. The entity-relationship (ER) data model developed by Chen [CHEN76] can be regarded to be the earliest semantic data model. In this model, the world is viewed as a collection of entities and relationships between entities. Figure 2-3 illustrates two entities, EMP and DEPT. The relationship between them is WORKS.

Relationships can be either one-one, many-one, or many-many. If it is assumed that each employee works in one department and each department has one employee, then WORKS is a one-one relationship. If it is assumed that an employee works in one department and each department can have many employees, then WORKS is a many-one relationship. If it is assumed that an employee works in many departments, and each department has many employees, then WORKS is a many-many relationship.

Figure 2-3. Entity-Relationship Representation

Note that various extensions to the entity-relationship model have been proposed. Also, ER models are used mainly to design databases.

2.2.4 Object-Oriented Data Model

With an object-oriented data model, the database is viewed as a collection of objects [BANE87]. Each object has a unique identifier called the object-ID. Objects with similar properties are grouped into a class. For example, employee objects are grouped into EMP class while department objects are grouped into DEPT class as shown in Figure 2-4. A class has instance variables describing the properties. Instance variables of EMP are SS#, Ename, Salary, and D#, while the instance variables of DEPT are D#, Dname, and Mgr. Objects in a class are its instances. As illustrated in the figure, EMP has three instances and DEPT has two instances.

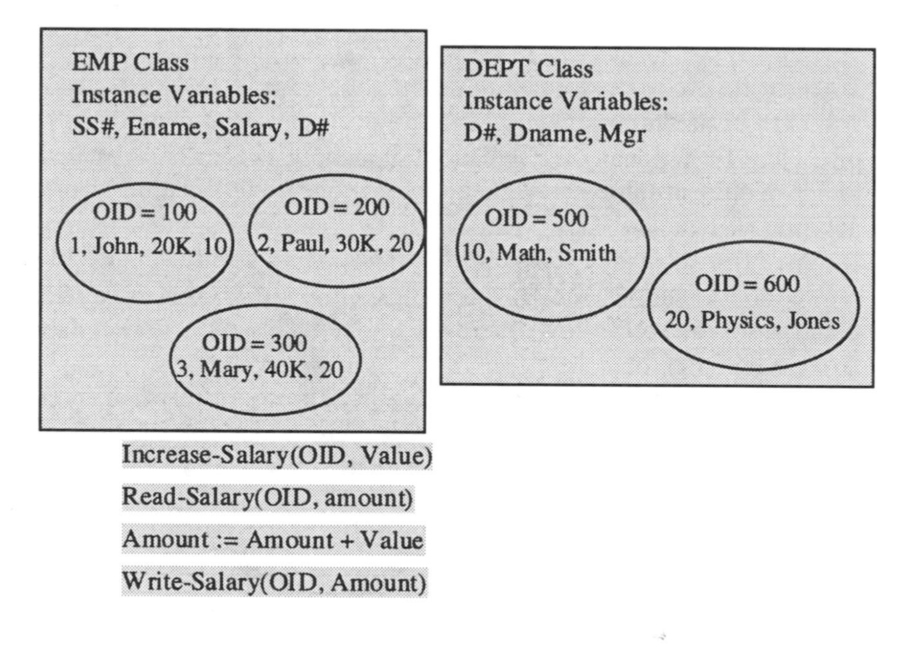

Figure 2-4. Objects and Classes

A key concept in object-oriented data modeling is encapsulation. That is, an object has well-defined interfaces. The state of an object can only be accessed through interface procedures called methods. For example, EMP may have a method called Increase-Salary. The code for Increase-Salary is illustrated in figure 2-4. A message, say Increase-Salary(1, 10K), may be sent to the object with object ID of 1. The object's current salary is read and updated by 10K. A second key concept in an object model is inheritance where a subclass inherits properties from its parent class. This feature is illustrated in Figure 2-5, where the EMP class has MGR (manager) and ENG (engineer) as its subclasses. Other key concepts in an object model include polymorphism and aggregation.[5] These features are discussed in [BANE87]. Further information can also be obtained in [THUR97].

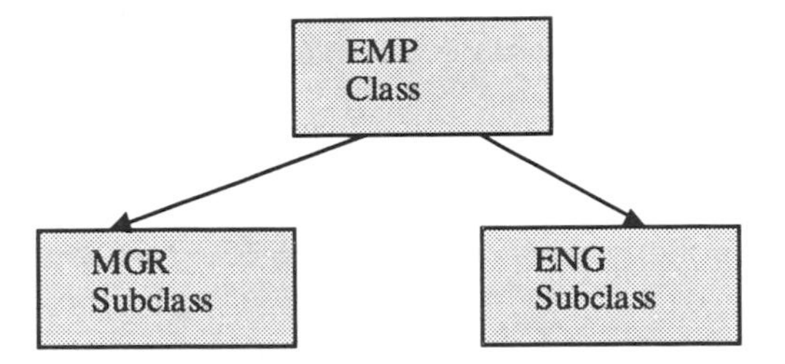

Figure 2-5. Class-Subclass Hierarchy

2.2.5 Object-Relational Data Model

Object-relational database systems were developed to overcome some of the problems with relational and object-oriented database systems. Relational data model is based on well-defined principles. Furthermore, a notable feature of relational database systems is the query language. The SQL language, developed initially for relational databases, is an ANSI standard. However, relational data models cannot support complex objects which are needed for new generation applications such as CAD/CAM and multimedia. On the other hand, object-oriented data models can support complex structures. However, in general, object-oriented database systems do not have good support for querying.

Book Extended Relation

ISBN#	Components
1	-
2	- - -
3	- - -

Figure 2-6. Object-Relational Model

5 Inheritance is also known as the IS-A hierarchy. Aggregation is also known as the IS-PART-OF hierarchy.

To overcome these problems, relational database vendors are building some sort of support for objects. Object-oriented database vendors are developing better query interfaces as well as better support to represent relationships. In addition, a third kind of system, object-relational database system, has been developed. These systems provide support both for relations and objects. Note that there is no standard object-relational model. With one approach, the relations are extended so that the data elements are no longer atomic. That is, the data elements could be complex objects. Figure 2-6 illustrates this concept where the book relation has an attribute called components. This attribute describes the components of the book. Object-relational systems are still young and we can expect them to mature over the next few years. Several object-oriented database system products, as well as a few object-relational database system prototypes, are discussed in [ACM91].

2.2.6 Logic-Based Data Model

The last model we describe here is the logic-based data model. Such a model received prominence in the late 1970s after the logic and database workshop in France. It was at this time that logic programming languages like Prolog were becoming popular. However, much of the development with the logic model took place after the start of the Japanese Fifth Generation Project in the early 1980s when it was declared that logic programming was the desired language for the project.

```
Parent(John, Mary) <-
Parent(Mary, Jane) <-
Parent(Mary, Jim) <-
Grandparent(X,Z) <- Parent (X,Y) and Parent (Y,Z)
```

Figure 2-7. Logic for Representation

Essentially, a logic model views the database as a collection of logic clauses. Note that logic clauses could be based on first order logic or higher order logic. However, due to the tight integration with relational databases and logic programming, the clauses are generally based on a restricted first order logic called Horn clause logic. Figure 2-7 illustrates a database which uses Horn logic for data representation. It essentially describes the parent-grandparent database. It consists of three clauses describing the parent relationship and one clause describing a rule for grandparent relationship. This way one does not have to store all the data for the grandparent relation. From the parent relation and the rule, one could deduce the grandparent relation. Further details on DBMSs based on logic as a data model are given in Chapter 8. For a discussion of logic programming we refer to [LLOY87].

2.3 ARCHITECTURAL ISSUES

This section describes various types of architectures for a database system. First we illustrate a very high-level centralized architecture for a database system. Then we describe a functional architecture for a database system. In particular, the functions of the DBMS component of the database system are illustrated in this architecture. Then we discuss the ANSI/SPARC's three-schema architecture which has been more or less accepted by the database community [DATE90]. Finally, we describe extensible architectures.[6]

Figure 2-8 is an example of a centralized architecture. Here, the DBMS is a monolithic entity and manages a database which is centralized. Functional architecture illustrates the functional modules of a DBMS. The major modules of a DBMS include the query processor, transaction manager, metadata manager, storage manager, integrity manager, and security manager. The functional architecture of the DBMS component of the centralized database system architecture (of Figure 2-8) is illustrated in Figure 2-9.

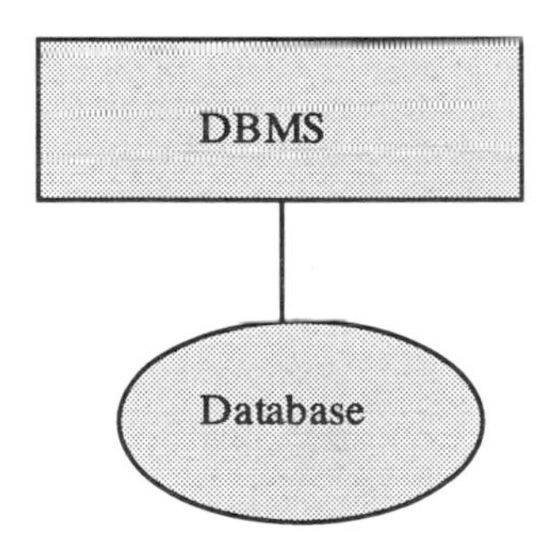

Figure 2-8. Centralized Architecture

[6] Note that distributed architectures for data management are discussed in Chapter 10 where we address data mining in distributed, heterogeneous, and legacy databases.

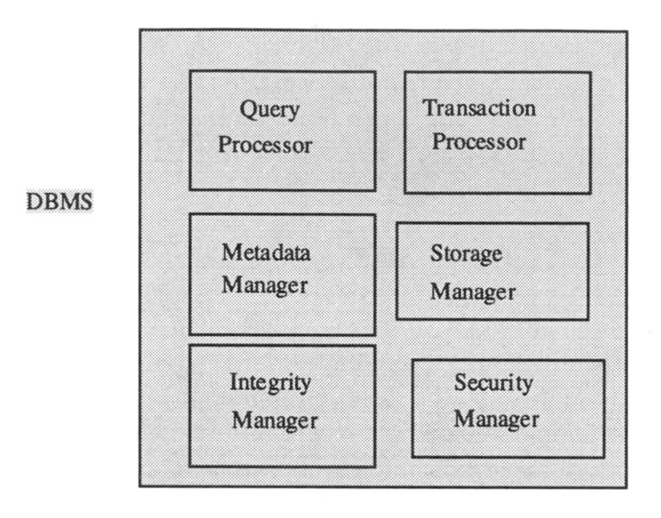

Figure 2-9. Functional Architecture for a DBMS

Schema describes the data in the database. It has also been referred to as the data dictionary or contents of the metadatabase. Three-schema architecture was proposed for a centralized database system in the 1960s. This is illustrated in Figure 2-10. The levels are the external schema which provides an external view, the conceptual schema which provides a conceptual view, and the internal schema which provides an internal view. Mappings between the different schemas must be provided to transform one representation into another. For example, at the external level, one could use ER representation. At the logical or conceptual level, one could use relational representation. At the physical level, one could use a representation based on B-Trees.[7]

There is also another aspect to architectures, and that is extensible database architectures. For example, for many applications, a DBMS may have to be extended with a layer to support objects or to process rules or to handle multimedia data types or even to do mining. Such an extensible architecture is illustrated in Figure 2-11.

---

7 Note that a B-Tree is a representation scheme used to physically represent the data. However, it is at a higher level than the bits and bytes level. For a discussion on physical structures and models, we refer to [DATE90].

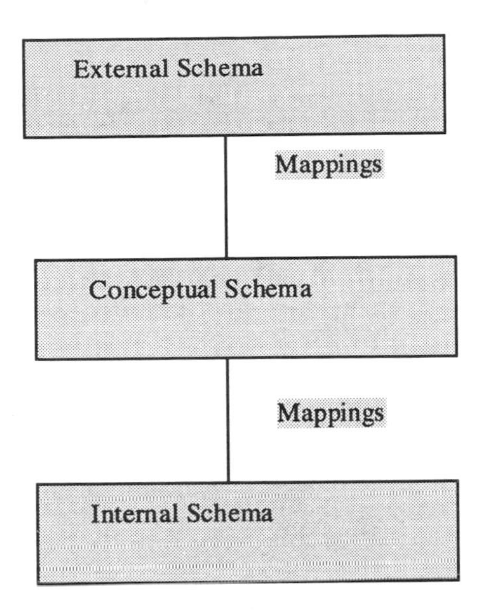

Figure 2-10. Three-Schema Architecture

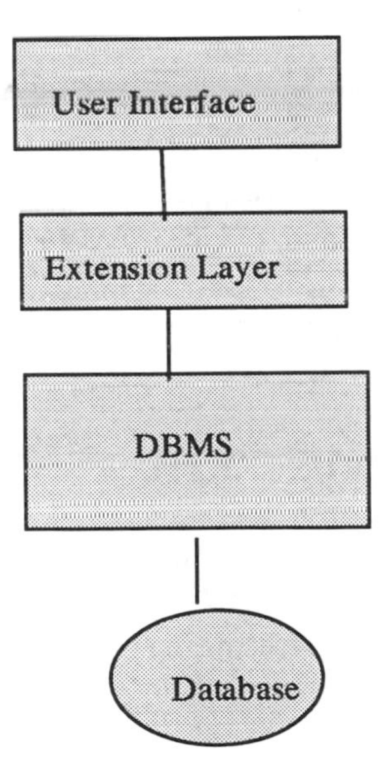

Figure 2-11. Extensible DBMS

2.4 DATABASE DESIGN

Designing a database is a complex process. Much of the work has been on designing relational databases. There are three steps which are illustrated in Figure 2-12. The first step is to capture the entities of the application and the relationships between the entities. One could use a model such as the entity-relationship model for this purpose. More recently, object-oriented data models, which are part of object-oriented design and analysis methodologies, are becoming popular to represent the application.

The second step is to generate the relations from the representations. For example, from the entity-relationship diagram of Figure 2-3, one could generate the relations EMP, DEPT, and WORKS. The relation WORKS will capture the relationship between employees and departments.

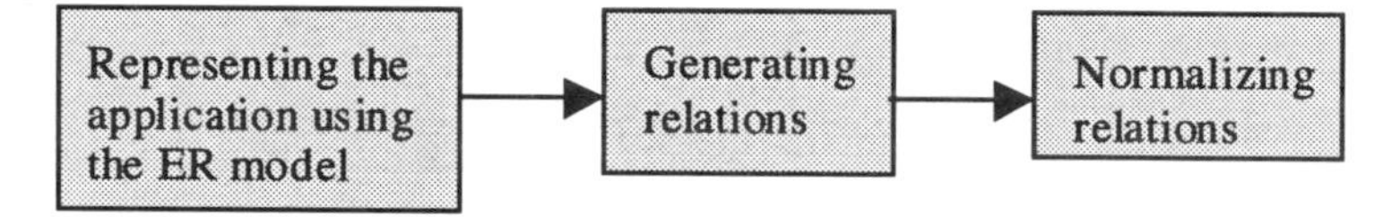

Figure 2-12. Database Design Process

The third step is to design good relations. This is the normalization process. Various normal forms have been defined in the literature (see, for example, [MAIE83] and DATE90]). For many applications, relations in third normal form would suffice. With this normal form, redundancies, complex values, and other situations that could cause potential anomalies are eliminated.

2.5 DATABASE ADMINISTRATION

A database has a database administrator (DBA). It is the responsibility of the DBA to define the various schemas and mappings. In addition, the functions of the administrator include auditing the database as well as implementing appropriate backup and recovery procedures.

The DBA could also be responsible for maintaining the security of the system. In some cases, security is maintained by the system security officer (SSO). The administrator should determine the granularity of the data for auditing. For example, in some cases there is tuple (or row) level auditing while in other cases there is table (or relation) level auditing. It is also the administrator's responsibility to analyze the audit data.

Note that there is a difference between database administration and data administration. Database administration assumes there is an installed database system. The DBA manages this system. Data administration functions include conducting data analysis, determining how a corporation handles its data, and enforcing appropriate policies and procedures for managing the data of a corporation. Data administration functions are carried out by the data administrator. For a discussion of data administration, we refer to [DMH94. DMH95, DOD94, DOD95]. Figure 2-13 illustrates various database administration issues.

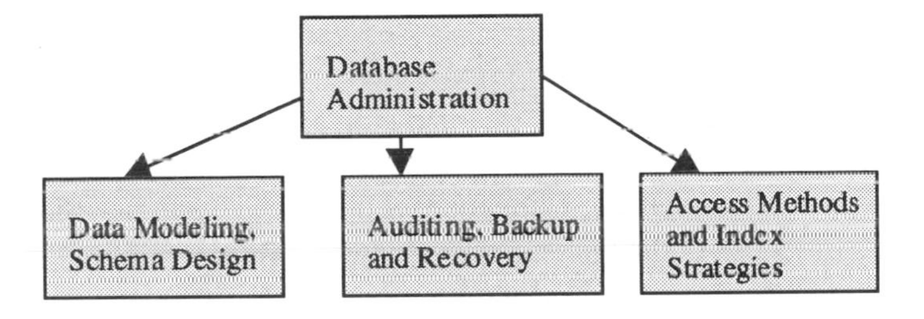

Figure 2-13. Some Database Administration Issues

2.6 DATABASE MANAGEMENT SYSTEM FUNCTIONS

2.6.1 Overview

The functional architecture of a DBMS was illustrated in Figure 2-9. The functions of a DBMS carry out its operations. A DBMS essentially manages a database, and it provides support to the user by enabling him to query and update the database. Therefore, the basic functions of a DBMS are query processing and update processing. In some applications such as banking, queries and updates are issued as part of transactions. Therefore transaction management is also another function of a DBMS. To carry out these functions, information about the database has to be maintained. This information is called metadata. The function that is associated with managing the metadata is metadata management. Special techniques are needed to manage the data stores that actually store the data. The function that is associated with managing these techniques is storage management. To ensure that the above functions are carried out properly and that the user gets accurate data, there are some additional functions. These include security management, integrity management, and fault management (i.e., fault tolerance).

The above are some of the essential functions of a DBMS. However, more recently there is emphasis on extracting information from the data. Therefore, other functions of a DBMS may include providing support for data mining, data warehousing, and collaboration.

This section focuses only on the essential functions of a DBMS. These are: query processing, transaction management, metadata management, storage management, maintaining integrity, security control, and fault tolerance. Note that we do not have a special section for update processing, as we can handle it as part of transaction management. We discuss each of the essential functions in Sections 2.6.2 to 2.6.7.

2.6.2 Query Processing

Query operation is the most commonly used function in a DBMS. It should be possible for users to query the database and obtain answers to their queries. There are several aspects to query processing. First of all, a good query language is needed. Languages such as SQL are popular for relational databases. Such languages are being extended for other types of databases. The second aspect is techniques for query processing. Numerous algorithms have been proposed for query processing in general and for the JOIN operation in particular (see, also [KIM85]). Also, different strategies are possible to execute a particular query. The costs for the various strategies are computed, and the one with the least cost is usually selected for processing. This process is called query optimization. Cost is generally determined by the disk access. The goal is to minimize disk access in processing a query.

As stated earlier, users pose a query using a language. The constructs of the language have to be transformed into the constructs understood by the database system. This process is called query transformation. Query transformation is carried out in stages based on the various schemas. For example, a query based on the external schema is transformed into a query on the conceptual schema. This is then transformed into a query on the physical schema. In general, rules used in the transformation process include the factoring of common subexpressions and pushing selections and projections down in the query tree as much as possible. If selections and projections are performed before the joins, then the cost of the joins can be reduced by a considerable amount.

Figure 2-14 illustrates the modules in query processing. The user interface manager accepts queries, parses the queries, then gives them to the query transformer. The query transformer and query optimizer communicate with each other to produce an execution strategy. The database is accessed through the storage manager. Responses are given to the user by the response manager.

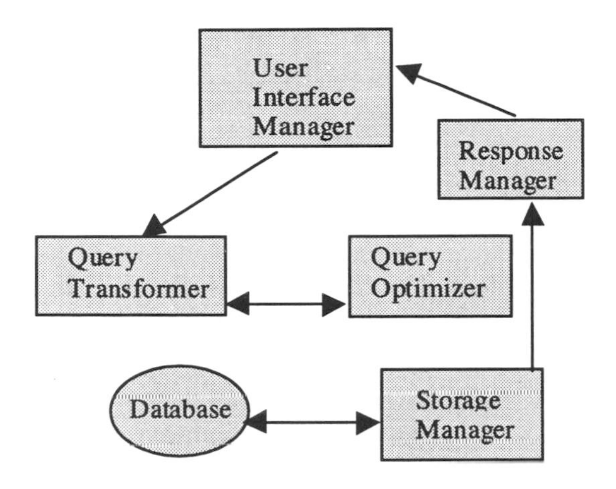

Figure 2-14. Query Processor

2.6.3 Transaction Management

A transaction is a program unit that must be executed in its entirety or not executed at all. If transactions are executed serially, then there is a performance bottleneck. Therefore, transactions are executed concurrently. Appropriate techniques must ensure that the database is consistent when multiple transactions update the database. That is, transactions must satisfy the ACID (Atomicity, Consistency, Isolation, and Durability) properties. Major aspects of transaction management are serializability, concurrency control, and recovery. We discuss them briefly in this section. For a detailed discussion of transaction management we refer to [DATE90] and ULLM88]. A good theoretical treatment of this topic is given in [BERN87].

Serializability: A schedule is a sequence of operations performed by multiple transactions. Two schedules are equivalent if their outcomes are the same. A serial schedule is a schedule where no two transactions execute concurrently. An objective in transaction management is to ensure that any schedule is equivalent to a serial schedule. Such a schedule is called a serializable schedule. Various conditions for testing the serializability of a schedule have been formulated for a DBMS.

Concurrency Control: Concurrency control techniques ensure that the database is in a consistent state when multiple transactions update the database. Three popular concurrency control techniques which ensure the serializability of schedules are locking, time-stamping, and validation.

Recovery: If a transaction aborts due to some failure, then the database must be brought to a consistent state. This is transaction recovery. One solution to handling transaction failure is to maintain log files. The transaction's actions are recorded in the log file. So, if a transaction aborts, then the database is brought back to a consistent state by undoing the actions of the transaction. The information for the undo operation is found in the log file. Another solution is to record the actions of a transaction but not make any changes to the database. Only if a transaction commits should the database be updated. There are some issues, however. For example, the log files have to be kept in stable storage. Various modifications to the above techniques have been proposed to handle the different situations.

When transactions are executed at multiple data sources, then a protocol called two-phase commit is used to ensure that the multiple data sources are consistent. Figure 2-15 illustrates the various aspects of transaction management.

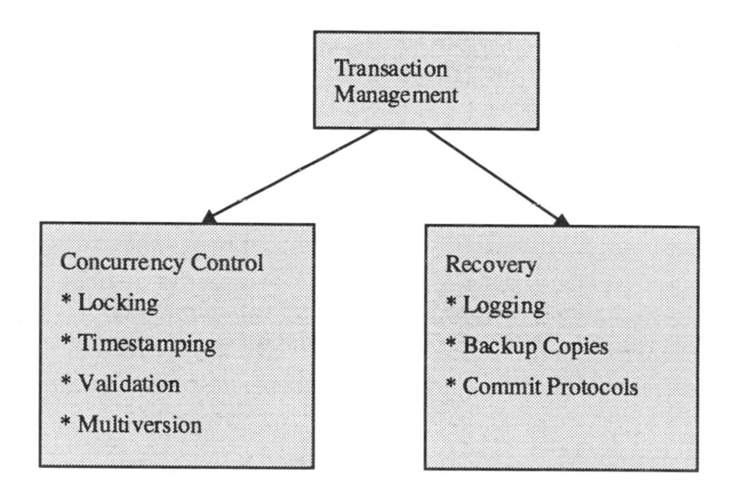

Figure 2-15. Some Aspects of Transaction Management

2.6.4 Storage Management

The storage manager is responsible for accessing the database. To improve the efficiency of query and update algorithms, appropriate access methods and index strategies have to be enforced. That is, in generating strategies for executing query and update requests, the access methods and index strategies that are used need to be taken into consideration. The access methods used to access the database would depend on the indexing methods. Therefore, creating and maintaining appropriate index files is a major issue in database management systems. By using an appropriate indexing mechanism, the query processing algorithms may not have to search the entire database. Instead, the data to be retrieved could be accessed directly. Consequently, the retrieval algorithms are more efficient. Figure 2-16 illustrates an example of an indexing strategy where the database is indexed by projects.

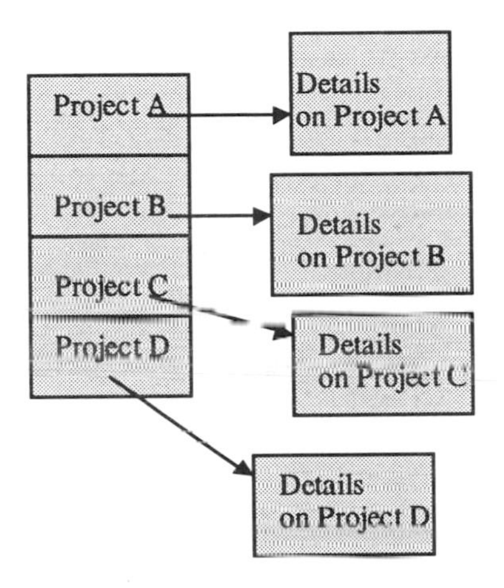

Figure 2-16. An Example Index on Projects

Much research has been carried out on developing appropriate access methods and index strategies for relational database systems. Some examples of index strategies are B-Trees and Hashing [DATE90]. Current research is focusing on developing such mechanisms for object-oriented database systems with support for multimedia data.

2.6.5 Metadata Management

Metadata describes the data in the database. For example, in the case of the relational database illustrated in Figure 2-2, metadata would include the following information: the database has two relations, EMP and DEPT; EMP has four attributes and DEPT has three attributes, etc. One of the main issues is developing a data model for metadata. In our example, one could use a relational model to model the metadata also. The metadata relation REL shown in Figure 2-17 consists of information about relations and attributes.

In addition to information about the data in the database, metadata also includes information on access methods, index strategies, security constraints, and integrity constraints. One could also include policies and procedures as part of the metadata. In other words, there is no standard definition for metadata. There are, however, efforts to standardize metadata [META96]. Metadata becomes a major issue with some of the recent developments in data management such as digital libraries. Some of the issues are discussed in Part III of this book.

Relation REL

Relation	Attribute
EMP	SS#
EMP	Ename
EMP	Salary
EMP	D#
DEPT	D#
DEPT	Dname
DEPT	Mgr

Figure 2-17. Metadata Relation

Once the metadata is defined, the issues include managing the metadata. What are the techniques for querying and updating the metadata? Since all of the other DBMS components need to access the metadata for processing, what are the interfaces between the metadata manager and the other components? Metadata management is fairly well understood for relational database systems. The current challenge is in managing the metadata for more complex systems such as digital libraries and Internet database systems.

2.6.6 Database Integrity

Concurrency control and recovery techniques maintain the integrity of the database. In addition, there is another type of database integrity and that is enforcing integrity constraints. There are two types of integrity constraints enforced in database systems. These are application independent integrity constraints and application specific integrity constraints. Integrity mechanisms also include techniques for determining the quality of the data. For example, what is the accuracy of the data and that of the source? What are the mechanisms for maintaining the quality of the data? How accurate is the data on output?

In [THUR97] we only discussed the enforcement of application independent and application specific integrity constraints. Our focus was on the relational data model. For a discussion of integrity based on data quality, we refer to [MIT]. Note that data quality is very important for mining and warehousing. If the data that is mined is not good, then one cannot rely on the results. We revisit this in Chapter 6.

Application independent integrity constraints include the primary key constraint, the entity integrity rule, referential integrity constraint, and the various functional dependencies involved in the normalization process (see the discussion in [DATE90]).

Application specific integrity constraints are those constraints that are specific to an application. Examples include "an employee's salary cannot decrease" and "no manager can manage more than two departments." Various techniques have been proposed to enforce application specific integrity constraints. For example, when the database is updated, these constraints are checked and the data is validated. Aspects of database integrity are illustrated in Figure 2-18.

2.6.7 Database Security

In this section we focus on discretionary security since this is the area that we are interested in with respect to warehousing and mining.[8] The major issues in security are authentication, identification, and enforcing appropriate access controls. For example, what are the mechanisms for identifying and authenticating the user? Will simple password mechanisms suffice? With respect to access control rules, languages such as SQL have incorporated GRANT and REVOKE statements to grant and revoke access to users. For many applications simple GRANT and REVOKE statements are not sufficient. There may be more complex authorizations based on database content. Negative authorizations may also be needed. Access to data based on the roles of the user is also being investigated.

Numerous papers have been published on discretionary security in databases. These can be found in various security related journals and conference proceedings (see, for example, [IFIP]). Some aspects of database security are illustrated in Figure 2-19.

---

[8] Note that multilevel security issues for database systems were addressed in [THUR97].

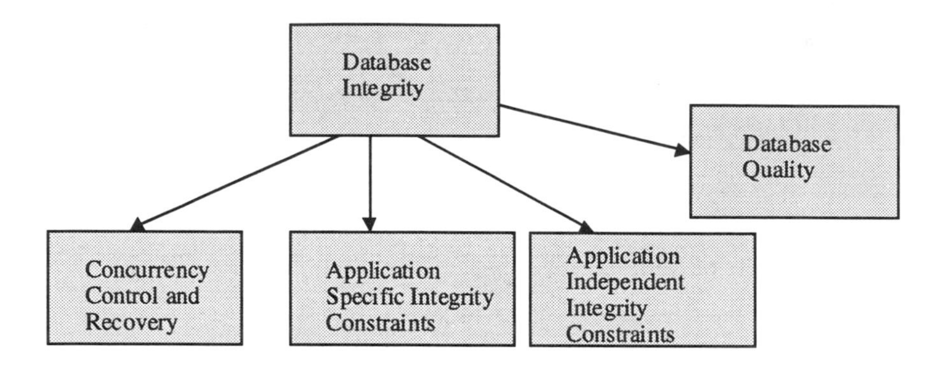

Figure 2-18. Some Aspects of Database Integrity

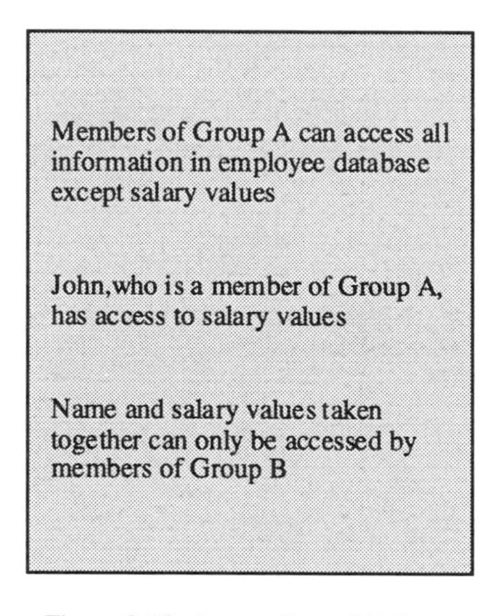

Figure 2-19. Access Control Rules

2.6.8 Fault Tolerance

The previous two sections discussed database integrity and security. A closely related feature is fault tolerance. It is almost impossible to guarantee that the database will function as planned. In reality, various faults could occur. These could be hardware faults or software faults. As mentioned earlier, one of the major issues in transaction management is to ensure that the database is brought back to a consistent state in the presence of faults. The solutions proposed include maintaining appropriate log files to record the actions of a transaction in case its actions have to be retraced.

Another approach to handling faults is checkpointing. Various checkpoints are placed during the course of database processing. At each checkpoint it is ensured that the database is in a consistent state. Therefore, if a fault occurs during processing, then the database must be brought back to the last checkpoint. This way it can be guaranteed that the database is consistent. Closely associated with checkpointing are acceptance tests. After various processing steps, the acceptance tests are checked. If the techniques pass the tests, then they can proceed further. Some aspects of fault tolerance are illustrated in Figure 2-20.

Checkpoint A
Start Processing
*
*
Acceptance Test
If OK, then go to Checkpoint B
Else Roll Back to Checkpoint A

Checkpoint B
Start Processing
*
*

Figure 2-20. Some Aspects of Fault Tolerance

2.7 DATABASE SYSTEMS AND MINING

2.7.1 Overview

As we have stressed repeatedly, data is critical for mining. Therefore, we need database systems to effectively manage the data that can be mined. These systems could be data warehousing systems or the database systems we have discussed in this chapter. Since data warehousing is a major area for data mining, we have devoted an entire chapter to data warehousing and its relationships to mining. These aspects will be described in Chapter 3. In this section we discuss other database system aspects for mining.

In Section 2.7.2 we focus on architecture, data modeling, database design and administration issues. For example, should a data mining tool[9] be tightly integrated with a database system or should it be loosely integrated? What is the impact of data modeling on data mining? Can one design a database in a such a way to facilitate mining? What is the impact on administration functions? In Section 2.7.3 we discuss the impact of data mining on the various functions discussed in this chapter. These include query processing, transaction management, storage management, metadata management, integrity and data quality, security, and fault tolerance.

2.7.2 Architectural, Modeling, Design, and Administration Aspects

As we mentioned in Chapter 1, data mining techniques have been around for a while. That is, various statistical reasoning techniques neural network-based techniques and various other artificial intelligence techniques have been around for decades. So why then is data mining becoming ever so popular now? The main reason is that we now have the data to mine. Data is now being collected, organized and structured, and database systems have played a major role in this. That is, as described in the previous sections of this chapter, with database systems we can now represent the data, store and retrieve the data, and enforce features such as integrity and security. This section addresses data mining with respect to architectural, modeling, design, and administration features.

So now that we have the data stored in databases and perhaps normalized and structured, how can we mine the data? One approach is to augment a DBMS with a mining tool as illustrated in Figure 2-21. One can buy a commercial DBMS and a commercial mining tool that has interfaces built to the DBMS and apply the tool to the data managed by the DBMS. This way the tool does not have to be burdened with getting the data to be mined. While this approach has advantages and promotes open architectures, there are some drawbacks. There could be some performance problems when you use a general purpose DBMS for mining.

The other approach is a tight integration with mining tools as shown in Figure 2-22. The database engine has mining tools incorporated within it. One can call such a DBMS to be a Mining DBMS. This way the various DBMS functions such as query processing and storage management are impacted by the mining techniques. For example, the optimization algorithms can be impacted by the mining techniques. There is much research to integrate mining into the DBMS engine (see, for example, [TSUR98]).

---

[9] We will also call such data mining tools data miners.

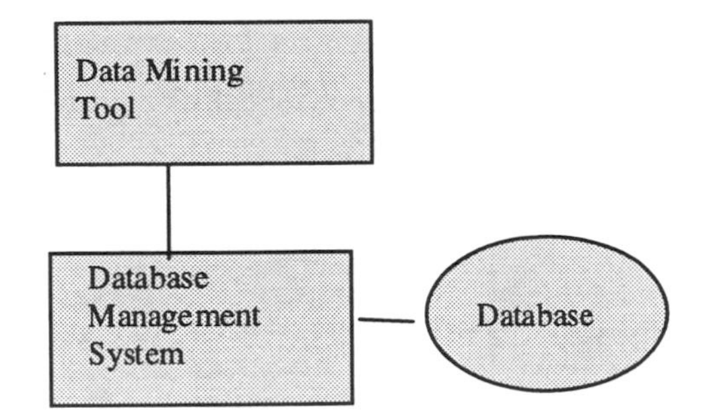

Figure 2-21. Loose Integration between DBMS and Data Miner

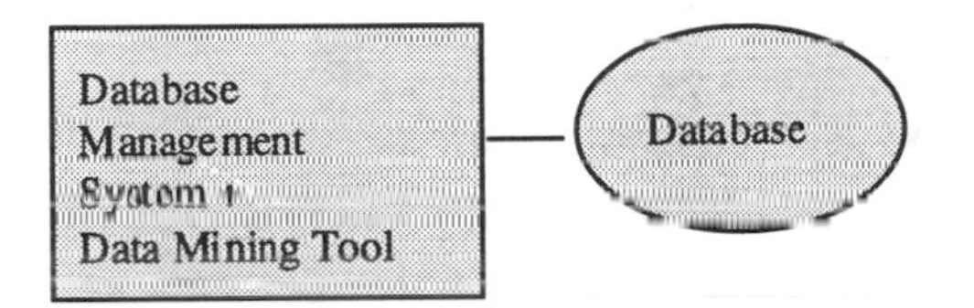

Figure 2-22. Tight Integration between DBMS and Data Miner

Mining DBMS also would mean eliminating unnecessary functions of a DBMS and focusing on key features. For example, transaction processing is a function supported by most commercial DBMSs. However, data mining is conducted usually not on transactional data but on decision support data. This data may not be data that is updated often by transactions. So, functions like transaction management could be removed from a Mining DBMS, and one could focus on additional features such as providing data integrity and quality. Note that there are cases where transactional data has to be mined, for instance, mining credit card transactions.

In general, in the case of a Mining DBMS, every function may be impacted by mining. These include query processing, storage manager, transaction manager, metadata manager, security and integrity manager. Therefore, we have added a data miner as part of a Mining DBMS and this is illustrated in Figure 2-23.

Next, let us focus on data modeling. The type of model used may have some impact on mining. Much of the data that is being currently is stored in relational databases. However, more and more data are now stored in nonrelational databases such as object-oriented, object-relational and multimedia databases. There is little information on mining object-based databases. There are some efforts on mining multimedia databases. This topic is discussed in Chapter 10. Figure

2-24 illustrates an example of mining an object database where the relationships between the objects are extracted first and stored in a relational database, and then the mining tools are applied to the relational database.

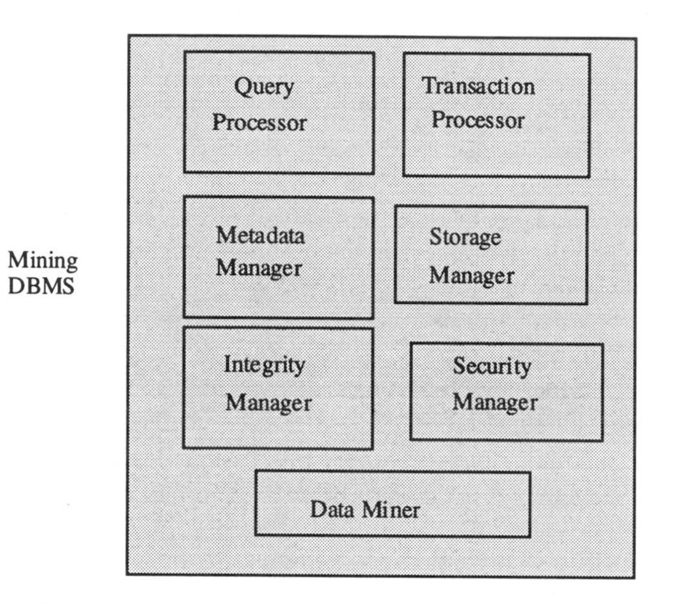

Figure 2-23. Functions of a Mining DBMS

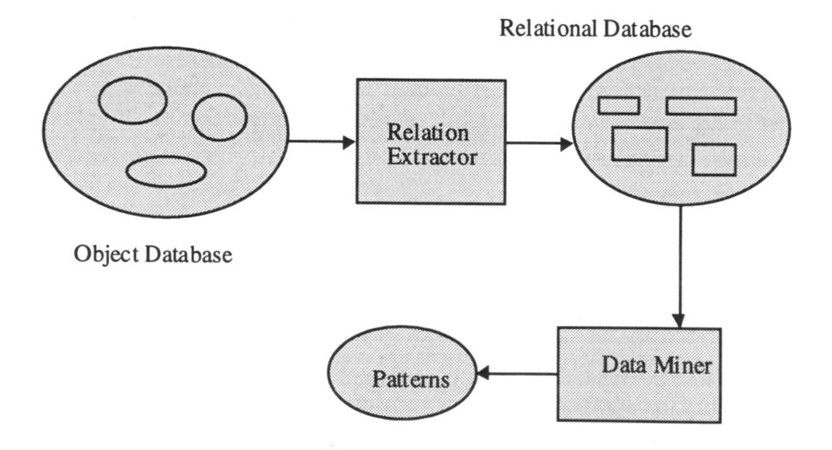

Figure 2-24. Mining Object Databases

Database design plays a major role in mining. For example, in the case of data warehousing, various approaches have been proposed to model and subsequently design the warehouse. These include multidimensional data models and on-line analytical processing models. Various schemas such as the star-schema have been proposed for data warehousing. As mentioned, organizing the data effectively is critical for mining. Therefore, such models and schemas are important for mining also.

Database administration is also impacted by mining. If one is to integrate mining with a DBMS, the questions include how often should the data in the databases be mined? Can mining be used to analyze the audit data? If the data is updated frequently, then how does it impact mining? These are some of the interesting questions, and we expect answers as more information is obtained about integrating mining with the DBMS functions.

2.7.3 Data Mining and Database Functions

Especially in the case of tight integration between the DBMS and the Data Miner, there is an impact on the various database systems functions. For example, consider query processing. There are efforts to examine query languages such as SQL and determine if extensions are needed to support mining (see for example [ACM96a]). If there are additional constructs and queries that are complex, then the query optimizer has to be adapted to handle such cases. Closely related to query optimization is efficient storage structures, indexes, and access methods. Special mechanisms may be needed to support data mining in the query process.

In the case of transaction management, as mentioned earlier, mining may have little impact, since mining is usually done on decision support data and not on transactional data. However, there are cases where transactional data are analyzed for anomalies such as credit card and telephone card anomalies. Some of us have been notified by our credit card or telephone companies about abnormal patterns in the usage. This is usually done by analyzing the transactional data. Such data could also be mined.

In the case of metadata, one could mine metadata to extract useful information in cases where the data cannot be analyzable. This may be the situation for unstructured data whose metadata may be structured. On the other hand, metadata could be a very useful resource for a data miner. Metadata could give additional information to help with the mining process.

Security, integrity, data quality, and fault tolerance are impacted by data mining. In the case of security, mining could pose a major threat to security and privacy. Privacy issues are discussed in Chapter 13. On the other hand, mining can be used to detect intrusions as well as to analyze audit data. In the case of auditing, the data to be mined is the large quantity of audit data. One may apply data mining tools to detect abnormal patterns. For example, suppose an employee makes an excessive number of trips to a particular country and this fact is known by posing some queries. The next query to pose is whether the employee has associations with certain people from that country. If the answer is positive, then the employee's behavior is flagged. The use of data mining for analyzing audit databases is illustrated in Figure 2-25.

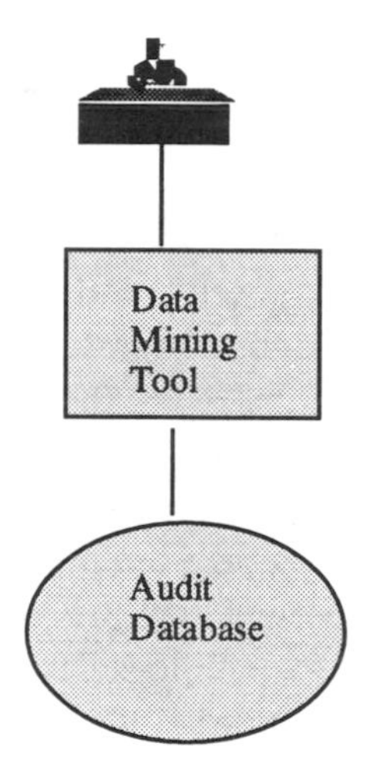

Figure 2-25. Mining an Audit Database

Note that data mining has many applications in intrusion detection and in analyzing threat databases. One can use data mining to detect patterns of intrusions and threats. This is an emerging area and is called Information Assurance. Not only is it important to have quality data, it is also important to recover from faults malicious or otherwise, and protect the data from threats or intrusions. While research in this area is just beginning, we can expect to see much progress.

In the case of data quality and integrity, one could apply mining techniques to detect bad data and improve the quality of the data. Mining can also be used to analyze safety data for various systems such as air traffic control systems, nuclear systems, and weapons systems. This is illustrated in Figure 2-26.

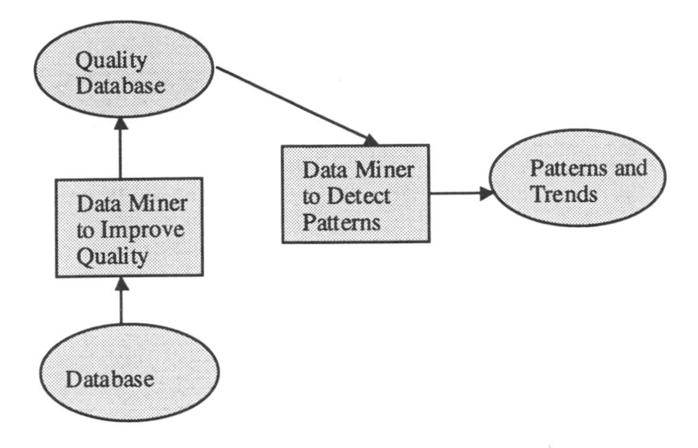

Figure 2-26. Data Quality and Data Mining

2.8 SUMMARY

This chapter has discussed various aspects of database systems and provided a lot of background information to understand the other chapters in this book. We began with a discussion of various data models. We chose relational, entity-relationship, object-oriented, object-relational, and logic-based models as they are most relevant to what we have addressed in this book. Then we provided an overview of various types of architectures for database systems. These include functional and schema architectures. Note that we have focused on centralized systems. Distributed database system issues will be discussed in a later chapter when we address mining distributed, heterogeneous and legacy databases. Next we discussed database design aspects and database administration issues. This chapter also provided an overview of the various functions of database systems. These include query processing, transaction management, storage management, metadata management, security, integrity, and fault tolerance. [10]

After providing an overview of database systems concepts, we then described the issues on integrating data mining with database systems. We discussed various architectures, both loose and tight coupling approaches. Then we discussed the impact of data modeling on mining such as mining relational databases and object-oriented databases. Next we discussed the impact of data mining on database design and data administration issues. Finally, data mining impact on the various database system functions such as query processing, transaction

---

10 Various texts have been published on database systems. Examples include [KORT86], [ULLM88], [DATE90], and [THUR97].

management, metadata management, security and integrity were discussed.

Many of the chapters in this book discuss various data management system aspects related to data mining. These include data warehousing, distributed and heterogeneous database systems, metadata management, multimedia databases, and security and privacy implications.

CHAPTER 3

DATA WAREHOUSING

3.1 OVERVIEW

Data warehousing is one of the key data management technologies to support data mining. Several organizations are building their own warehouses. Commercial database system vendors are marketing warehousing products. In addition, some companies are specializing only in developing data warehouses. What then is a data warehouse? The idea behind this is that it is often cumbersome to access data from the heterogeneous databases. Several processing modules need to cooperate with each other to process a query in a heterogeneous environment. Therefore, a data warehouse will bring together the essential data from the heterogeneous databases. This way the users need to query only the warehouse.

As stated by Inmon [INMO93], data warehouses are subject-oriented. Their design depends to a great extent on the application utilizing them. They integrate diverse and possibly heterogeneous data sources. They are persistent. That is, the warehouse is very much like a database. They vary with time. This is because as the data sources from which the warehouse is built get updated, the changes have to be reflected in the warehouse. Essentially data warehouses provide support to decision support functions of an enterprise or an organization. For example, while the data sources may have the raw data, the data warehouse may have correlated data, summary reports, and aggregate functions applied to the raw data.

Figure 3-1 illustrates a data warehouse. The data sources are managed by database systems A, B, and C. The information in these databases are merged and put into a warehouse. There are various ways to merge the information. One is to simply replicate the databases. This does not have any advantages over accessing the heterogeneous databases. The second case is to replicate the information, but to remove any inconsistencies and redundancies. This has some advantages as it is important to provide a consistent picture of the databases. The third approach is to select a subset of the information from the databases and place it in the warehouse. There are several issues here. How are the subsets selected? Are they selected at random or is some method used to select the data? For example, one could take every other row in a relation (assuming it is a relational database) and store these rows in the warehouse. The fourth approach, which is a slight variation of the third approach, is to determine the types of queries that users would pose, then analyze the data and store only the data that is required by the user. This is called on-line analytical processing (OLAP) as opposed to on-line transaction processing (OLTP).

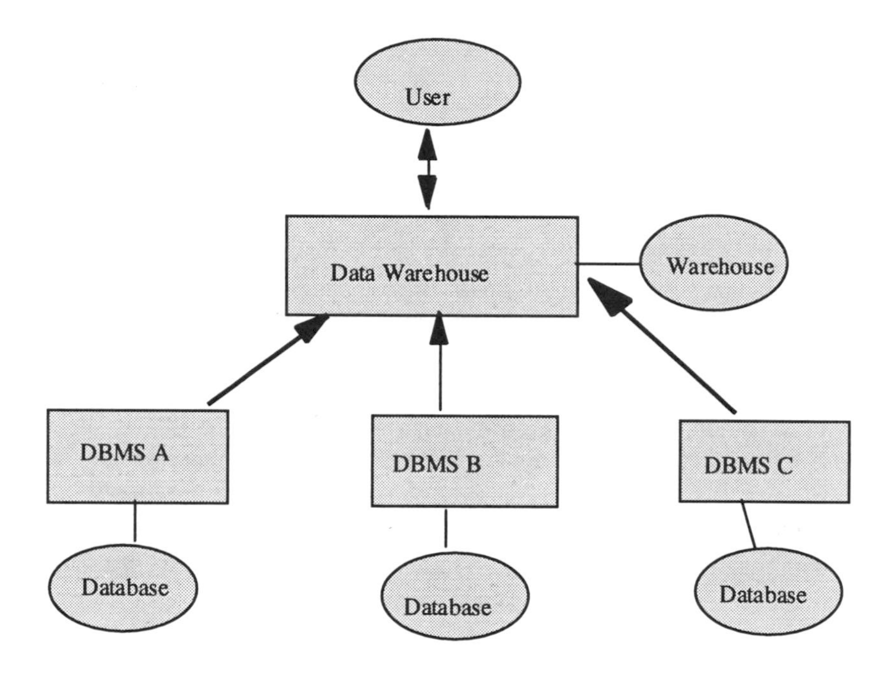

Figure 3-1. Data Warehouse Example

With a data warehouse, data may often be viewed differently by different applications. That is, the data is multidimensional. For example, the payroll department may want data to be in a certain format while the project department may want data to be in a different format. The warehouse must provide support for such multidimensional data. Multiple views of the same data are illustrated in Figure 3-2.

In integrating the data sources to form the warehouse, a challenge is to analyze the application and select appropriate data to be placed in the warehouse. At times, some computations may have to be performed so that only summaries and averages are stored in the data warehouse. Note that it is not always the case that the warehouse has all the information for a query. In this case, the warehouse may have to get the data from the heterogeneous data sources to complete the execution of the query. Another challenge is what happens to the warehouse when the individual databases are updated? How are the updates propagated to the warehouse? How can security be maintained? These are some of the issues that are being investigated.

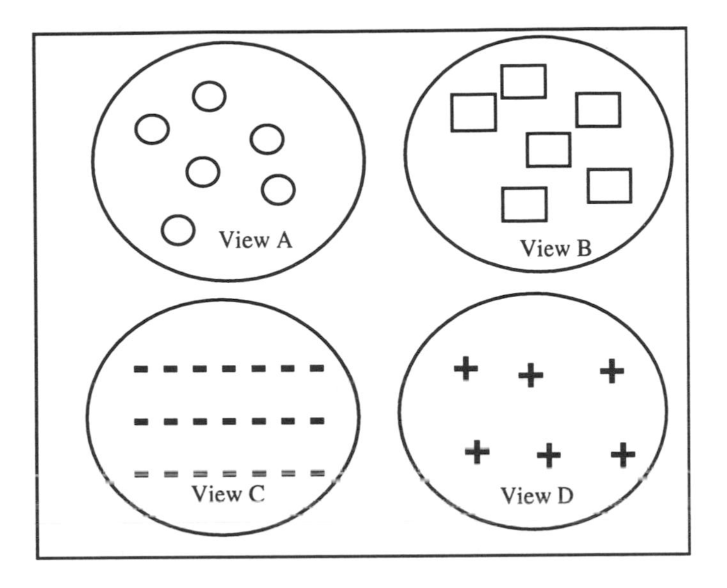

Figure 3-2. Multiple Views of the Data

In our tutorials on data warehousing and mining, we are often asked the question do we really want to build a warehouse or do we want to integrate the data sources as discussed in Chapter 10 of this book. For example, what is the difference between warehousing and interoperability? It should be noted that a warehouse is built for decision support. Therefore, the data in the warehouse may not reflect the changes made in the data sources in a timely fashion. If the data sources are changing rapidly and if the user wants to see the changes, then a warehouse may not make much sense, and one may want to simply integrate the heterogeneous data sources. However, in many cases one could have both; that is, a warehouse as well as database systems that interoperate with each other.

This chapter is organized in the following way. Technologies for data warehousing is the subject of Section 3.2. The discussion here includes the role of metadata as well as distributing the warehouse. Designing the data warehouse is discussed in Section 3.3. In particular, data modeling, data distribution, integrating heterogeneous data sources, and security are discussed. In Section 3.4, we describe data warehousing and its relationship to mining. The chapter is summarized in Section 3.5.

3.2 TECHNOLOGIES FOR DATA WAREHOUSING

Figure 3-3 illustrates data warehousing technologies. As can be seen in this figure, several technologies have to be integrated to develop a data warehouse. These include heterogeneous database integration, statistical databases, data modeling, metadata management, access methods and indexing, query language, database administration, database security, distributed database management, and high performance database management. In this section we briefly examine these technologies within the context of data warehousing.

Heterogeneous database integration is an essential component to data warehousing. This is because data from multiple heterogeneous data stores may have to be integrated to build the warehouse. There is, however, a major difference. Often in heterogeneous database integration there is no single repository to store the data. However, in a warehouse there is usually a single repository for the warehouse data and this repository has to be managed.

Statistical databases keep information such as sums, averages, and other aggregates. There are various issues for statistical databases. For example, how can summary data be maintained when the database gets updated? How can the individual data items be protected? For example, the average salary may be Unclassified while the individual salaries are Secret. Since warehouses keep summary information, techniques used to manage statistical databases need to be examined for warehouses.

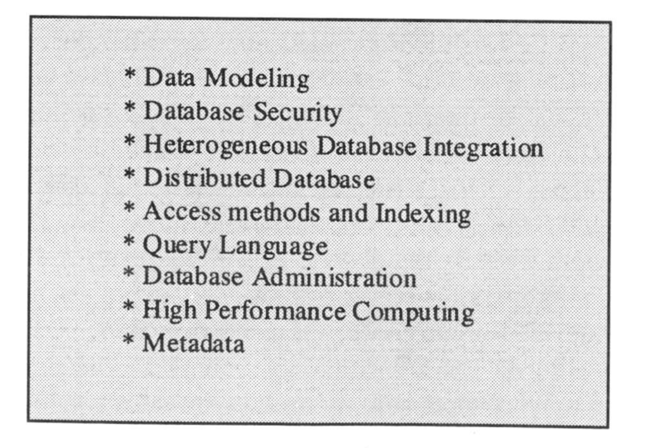

Figure 3-3. Some Data Warehousing Technologies

Data modeling is an essential task for building a data warehouse. Is the data model influenced by the data models used by the heterogeneous data sources? Should a data model be developed from scratch? Inmon has outlined several steps to developing a data model [INMO93]. He says that at the higher level there are three stages: developing a corporate model, an enterprise model, and a warehouse model. At the middle level there may be a model possibly for each subject, and at the physical level it includes features such as keys. Some argue this is too lengthy a process and that one should get to the warehouse model directly. As more experiences are reported on developing data warehouses, this issue may be resolved. New types of data models such as multidimensional data models and schemas such as star-schemas have been proposed for data warehousing. Essential points of the star-schema (also called *-schema) are illustrated in Figure 3-4. For example, in a project database, there is a central table that has key information on projects such as project number, project leader, estimated time duration, cost and other pertinent data.. Each of the entries in this table could be elaborated in other tables. For example, estimated time duration could be in days, months, and years. Cost could be dollars, pounds, yens and other currency. Depending on who is using the data, different views of the data could be provided to the user.

Appropriate access methods and index strategies have to be developed for the warehouse. For example, the warehouse is structured in such a way so as to facilitate query processing. An example query may be: how many red cars costing more than 50K were bought in 1995 by physicians? Many relations have to be joined to process this query. Instead of joining the actual data, one could get the result by combining the bit maps for the associated data. The warehouse may utilize an index strategy called a bit map index where essentially there is a 1 in the bit map if the answer is positive in the database. So, if the color of the car is red, then in the associated bit map, there will be a 1. This is a simple example. Current research is focusing on developing more complex access methods and index strategies.

Developing an appropriate query language for the warehouse is an issue. This would depend on the data model utilized. If the model is relational, then an SQL-based language may be appropriate. If the data model is object-oriented, then an ODMG-based language may be appropriate.[11] One may also need to provide visual interfaces for the warehouse.

---

[11] ODMG is the Object Database Management Group and is a consortium of about a dozen corporation, mainly from the object database community, specifying standards for object database management.

Database administration techniques may be utilized for administering the warehouse. Is there a warehouse administrator? What is the relationship between the warehouse administrator and the administrator of the data sources? How often should the warehouse be audited? Should the warehouse be audited? Inmon has given some reasons as to why it may not be a good idea to audit the warehouse [INMO93]. Another administration issue is propagating updates to the database. In many cases, the administrators of the data sources may not want to enforce triggers on their data. If this is the case, it may be difficult to automatically propagate the updates.

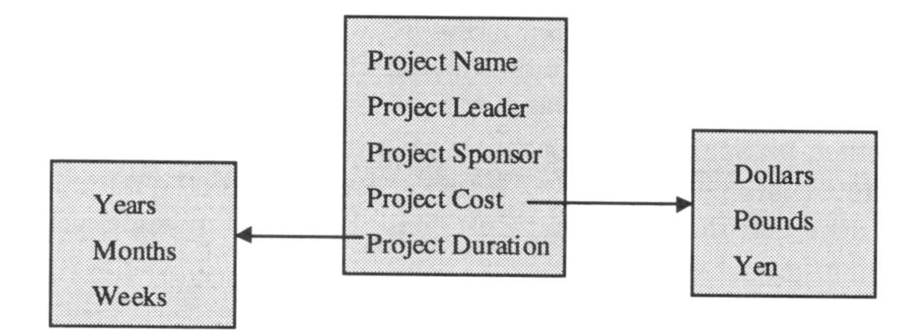

Figure 3-4. Multidimensional Tables

Protecting the warehouse is a major issue. Security issues for integrating heterogeneous database systems discussed in [THUR97] need to be examined here also. Furthermore, statistical database security also will play an import role. Security controls also have to be enforced in maintaining the warehouse. This will have an impact on querying, managing the metadata, and updating the warehouse. In addition, if multilevel security is needed, then there are additional considerations. For example, what are the trusted components of the warehouse?

High performance computing including parallel database management plays a major role in data warehousing. The goal is for users to get answers to complex queries rapidly. Therefore, parallel query processing strategies are becoming popular for warehouses. Appropriate hardware and software are needed for efficient query processing.

Metadata management is another critical technology for data warehousing. The problem is defining the metadata. Metadata could come from the data sources. Metadata will include the mappings between the data sources and the warehouse. There is also metadata specific to the warehouse. Many of the issues discussed in Chapter 2 for metadata management are applicable for the warehouse. Figure 3-5 illustrates the various types of metadata that must be maintained in developing and maintaining a warehouse. There are three types of metadata. One is metadata for the individual data sources. The second is the metadata needed for mappings and transformations to build the warehouse, and the third is the metadata to maintain and operate the warehouse.

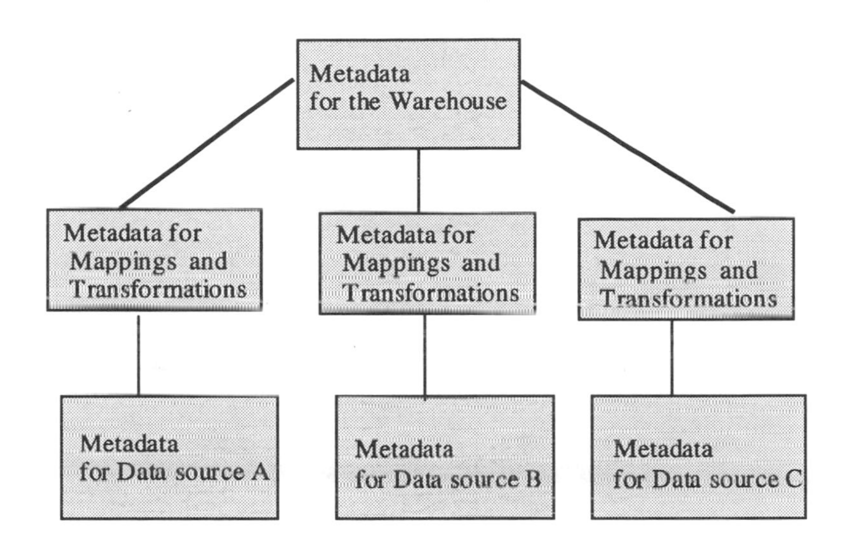

Figure 3-5. Types of Metadata

Distributed database technology plays a role in data warehousing. Should the warehouse be centralized or distributed? If it is distributed, then much of the technology for distributed database management discussed in Chapter 3 is applicable for data warehousing. Figures 3-6 and 3-7 illustrate architectures for nondistributed and distributed data warehouses. In the nondistributed case, there is a central warehouse for the multiple branches, say in a bank. In the distributed warehouse case, one may assume that each bank has its local warehouse and the warehouses communicate with each other.

3.3 DEVELOPING THE DATA WAREHOUSE

Designing and developing the data warehouse is a complex process and in many ways depends on the application. A good reference to data warehousing is the book by Inmon [INMO93]. It describes the details of the issues involved in building a data warehouse. In this section we outline some of the steps to designing the warehouse. Figure 3-8 illustrates some of these steps.

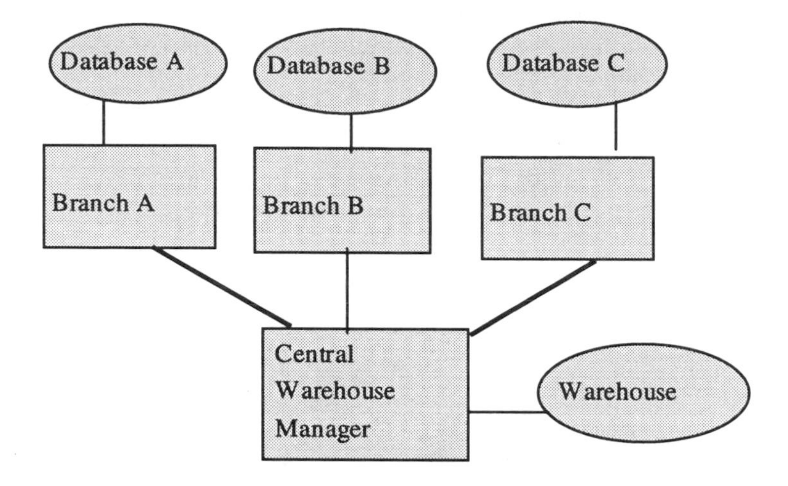

Figure 3-6. Nondistributed Data Warehouse

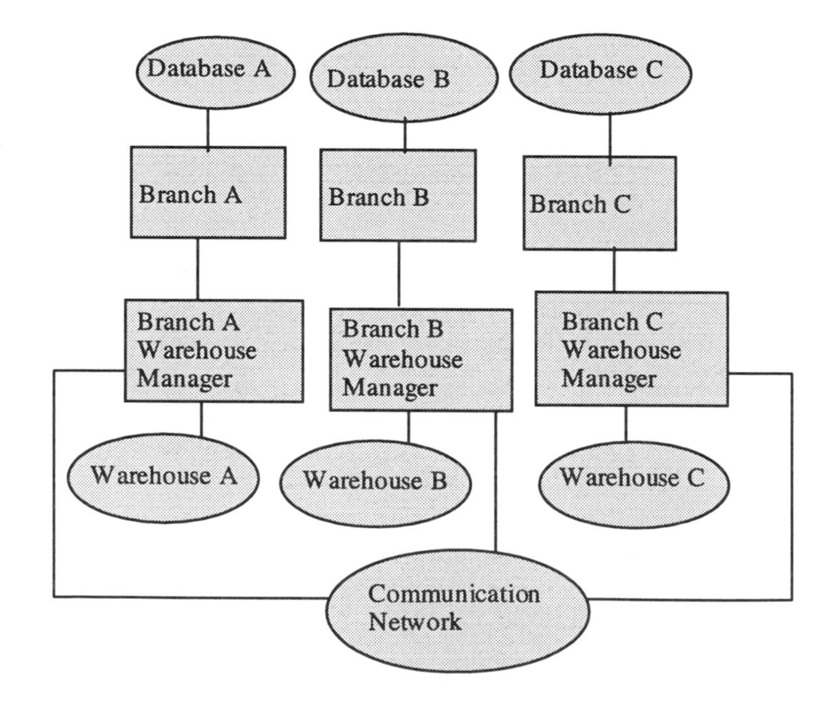

Figure 3-7. Distributed Data Warehouse

There are three phases to developing a warehouse. One phase focuses on structuring the warehouse so that query processing is facilitated. In other words, this phase focuses on getting the data out of the warehouse. Another phase focuses on bringing the data into the warehouse. For example, how can the heterogeneous data sources be integrated so that the data can be brought into the warehouse? The third phase maintains the warehouse once it is developed. This means the process does not end when the warehouse is developed. It has to be continually maintained. We first outline the steps in each of the phases.

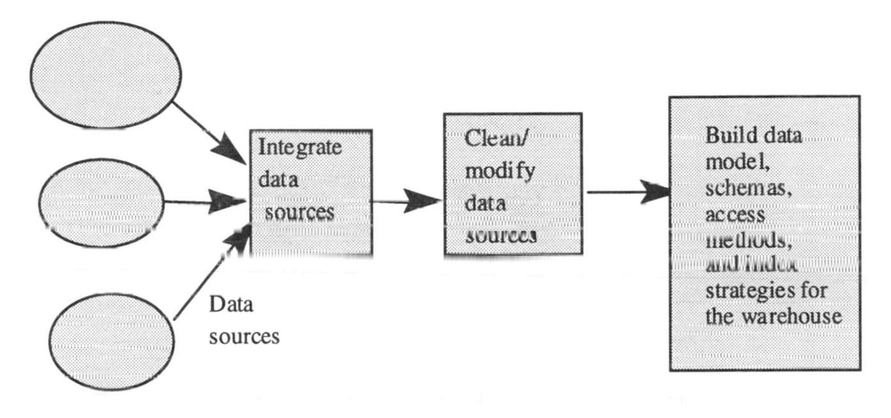

Figure 3-8. Developing a Data Warehouse

One of the key steps in getting the data out of the warehouse is application analysis. For example, what types of queries will the users pose? How often are the queries posed? Will the responses be straightforward? Will the users need information like summary reports? A list consisting of such questions needs to be formulated.

Another step is to determine what the user would expect from the warehouse. Would he want to deal with a relational model or an object-oriented model or both? Are multiple views needed? Once this is determined, how do you go about developing a data model? Are there intermediate models?

A third step is to determine the metadata, index strategies, and access methods. Once the query patterns and data models have been determined, one needs to determine what kinds of metadata have to be maintained. What are the index strategies and access methods enforced? What are the security controls?

A closely related task is developing the various schemas for the warehouse. Note that the individual databases will have their own schema. The complexity here is in integrating these schemas to develop a global schema for the warehouse. While schema integration techniques for distributed and heterogeneous databases may be used, the warehouse is developed mainly to answer specific queries for various applications. Therefore, special types of schemas such as star schemas and constellation schemas have been proposed in the literature. Products based on these schemas have also been developed. A discussion of these various types of schemas is beyond the scope of this book.

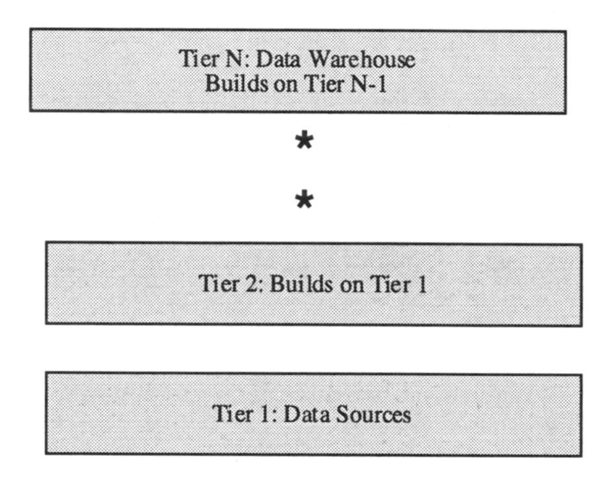

Figure 3-9. Multi-Tier Architecture

There are several technical issues in bringing the data into the warehouse from the different data sources. What information should be deleted from the individual databases when migrating the data to the warehouse? How should integrity be maintained? What is the security policy? How can inconsistencies be resolved? This requires a lot of work. Various algorithms for integrating heterogeneous databases have to be examined. At the end of this stage, one would have some form of a warehouse.[12] Multi-tier architecture is becoming popular for data warehousing (see, for example, [ROSE98]). Essentially, data passes through multiple tiers before reaching the warehouse. This is illustrated in Figure 3-9. At the bottom tier are the data sources. At the top tier is the data warehouse. Between the top and bottom there may be multiple tiers. Each tier has its own schemas, metadata, and various

---

[12] Note that we have also heard the term data marts used for the data stored in a warehouse that is relevant to a particular domain. We do not differentiate between data warehousing and data marts in this book.

administration details, and each tier takes advantage of the work done at lower tiers.

Once the warehouse is designed and developed, there are also some additional considerations for maintaining the warehouse. How is the security of the warehouse maintained? Should the warehouse be audited? How often is the warehouse updated? How are the changes to the local databases to be propagated to the warehouse? What happens if a user's query cannot be answered by the warehouse? Should the warehouse go to the individual databases to get the data if needed? How can data quality and integrity be maintained?

We have outlined a number of phases and steps to developing a data warehouse. The question is, should these phases and steps be carried out one after the other or should they be done in parallel? Based on discussions with those who have actually built warehouses, it seems that many of the activities can be done in parallel. As in most software systems, there is a planning phase, a development phase, and a maintenance phase. However, there are some additional complexities. The databases themselves may be migrating to new architectures or data models. This would have some impact on the warehouse. New databases may be added to the heterogeneous environment. The additional information should be migrated to the warehouse without causing inconsistencies. These are difficult problems and there are investigations on how to resolve them. Although there is much promise, there is a long way to go before viable commercial products are developed.

In summary, a data warehouse enables different applications to view the data differently. That is, it supports multidimensional data. Data warehouse technology is an integration of multiple technologies including heterogeneous database integration, statistical databases, and parallel processing. The challenges in data warehousing include developing appropriate data models, architectures (e.g., centralized or distributed), query languages, and access methods/index strategies, as well as developing techniques for query processing, metadata management, maintaining integrity and security, and integrating heterogeneous data sources. Integrating structured and unstructured databases, such as relational and multimedia databases, is also a challenge.

While the notion of data warehousing has been around for a while, it is only recently that we are seeing the emergence of commercial products. This is because many of the related technologies such as parallel processing, heterogeneous database integration, statistical databases, and data modeling have evolved a great deal and some of them are fairly mature technologies. There are now viable technologies to build a data warehouse. We expect the demand for data warehousing to grow rapidly over the next few years.

It should be noted that many of the developments in data warehousing focus on integrating data stored in structured databases such as relational databases. In the future we can expect to see multimedia data sources being integrated to form a warehouse.

3.4 DATA WAREHOUSING AND DATA MINING

Data warehousing has been the subject of discussion in this chapter. A data warehouse assembles the data from heterogeneous databases so that users query only a single point. The responses that a user gets to a query depend on the contents of the data warehouse. The data warehouse in general does not attempt to extract information from the data in the warehouse. While data warehousing formats the data and organizes the data to support management functions, data mining attempts to extract useful information as well as predicts trends from the data. Figure 3-10 illustrates the relationship between data warehousing and data mining. Note that having a warehouse is not necessary to do mining, as data mining can be applied to databases also. However, a warehouse structures the data in such a way as to facilitate mining, so in many cases it is highly desirable to have a data warehouse to carry out mining. The relationship between warehousing, mining and database systems is illustrated in Figure 3-11.

When giving various tutorials on data warehousing and mining, we are often asked the question where does warehousing end and mining begin? For example, is there a clear difference between warehousing and mining? This answer we think is subjective. There are certain questions that warehouses can answer. Furthermore, warehouses have built in decision support capabilities. Some warehouses carry out predictions and trends. In this case warehouses carry out some of the data mining functions. In general, we believe that in the case of a warehouse the answer is in the database. The warehouse has to come up with query optimization and access techniques to get the answer. For example, consider questions like "how many red cars did physicians buy in 1990 in New York?" The answer is in the database. However, for a question like "how many red cars do you think physicians will buy in 2005 in New York?" the answer may not be in the database. Based on the buying patterns of physicians in New York and their salary projections, one could predict the answer to this question.

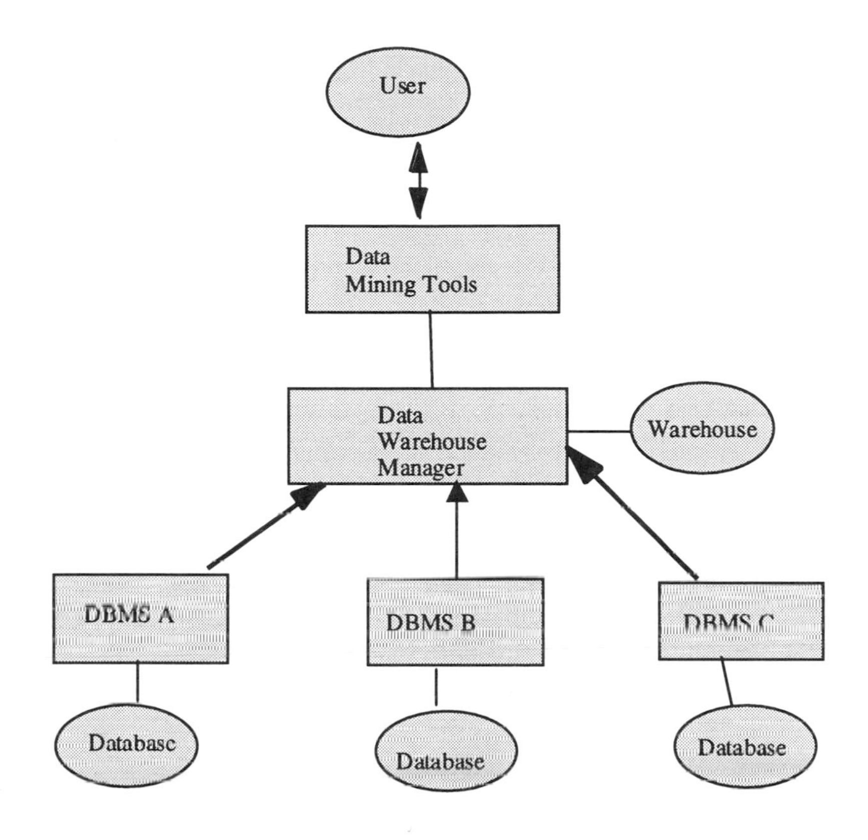

Figure 3-10. Data Mining vs. Data Warehousing

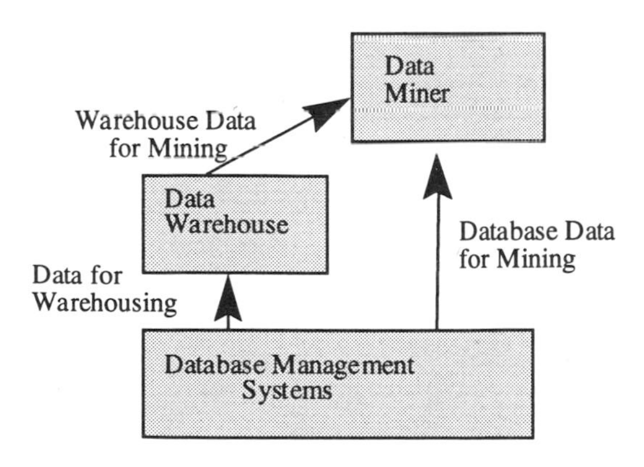

Figure 3-11. Database Systems, Data Warehousing and Mining

Essentially, a warehouse organizes the data effectively so the data can be mined. The question then is do you absolutely have to have a warehouse to mine the data? The answer we give is that it is very good to have a warehouse, but it does not mean we must have a warehouse to mine. A good DBMS that manages a database effectively could also be used. Also, with a warehouse one often does not have transactional data. Furthermore, the data may not be current, therefore the results obtained from mining may not be current. If one needs up-to-date information, then one could mine the database managed by a DBMS which also has transaction processing features. Mining data that keeps changing often is a challenge. Typically mining has been used for decision support data. Therefore, there are several issues that need further investigation before we can carry out what we call real-time data mining. For now at least, we believe that having a good data warehouse is critical to do good mining for decision support functions. Note that one could also have an integrated tool that carries out both data warehousing and data mining functions. We call such a tool a data warehouse miner and this is illustrated in Figure 3-12.

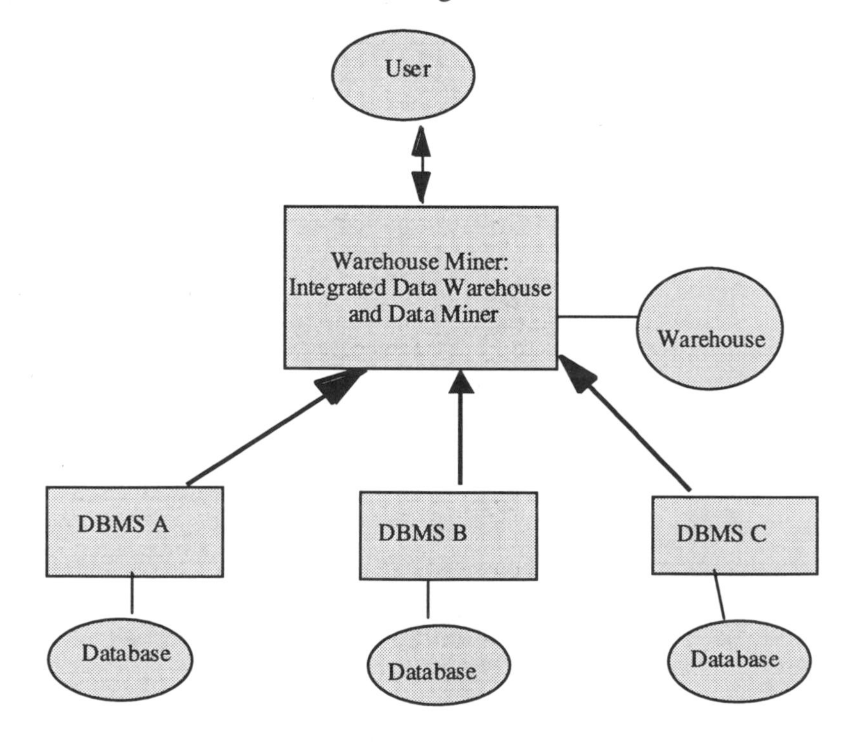

Figure 3-12. Integrated Data Warehousing and Data Mining

3.5 SUMMARY

This chapter addressed a key technology for data mining and that is data warehousing. We started with a discussion of a definition for a data warehouse, the technologies for a data warehouse, functions of a data warehouse, and issues on developing a data warehouse. Key concepts in warehousing include developing a data model, architecture, and access methods and index strategies for warehousing.

We then discussed the relationship between warehousing and mining. As mentioned earlier, a frequent question we are asked is where does warehousing end and mining begin. We discussed our answer to this question. We also discussed the need for warehousing to do mining. Finally we discussed the relationships between warehousing, database management, and mining.

As in the case of the relationship between a database system and a data miner, one could also have both tight and loose integration between a data warehouse and a data miner. In the case of loose integration, the data miner and warehouse could be developed by two different vendors. There has to be well-defined interfaces to integrate the miner with the warehouse. In the case of tight integration, a warehouse has to be integrated with the miner. In this case the query algorithms, access methods, and index strategies of the warehouse will be impacted by mining.

There are quite a few data warehousing products now on the market. They are from general purpose database system vendors as well as from special purpose warehouse vendors. In addition, several operational warehouse systems have been reported. Some of the warehousing products also integrate the data mining capabilities. We expect that there will be a large market for integrating warehousing with mining products.

CHAPTER 4

SOME OTHER SUPPORTING TECHNOLOGIES

4.1 OVERVIEW

In Chapters 2 and 3 we have taken a data-oriented perspective to data mining. This is for two reasons. First, much of our work has focused on data warehousing and in providing the support to data mining in organizing and structuring the data. Again our belief is that unless you have good data there is no point in doing mining. Furthermore, the other data mining technologies have been around for a while and some argue that not many of them have worked on real-world problems. However, integration of these technologies with data management has contributed a lot to data mining.

While we feel that data is important, if not for the other supporting data mining technologies we would not have data mining either. So this chapter provides a brief introduction to these other technologies and discusses how they support data mining. Notable among these technologies are statistical methods and machine learning. Statistical methods have resulted in various statistical packages to compute sums, averages, and distributions. These packages are now being integrated with databases for mining. Machine learning is all about learning rules and patterns from the data. One needs some amount of statistics to carry out machine learning. While statistical methods and machine learning are the two key components to data mining apart from data management, there are also some other technologies. These include visualization, parallel processing, and decision support. Visualization techniques help visualize the data so that data mining is facilitated. Parallel processing techniques help improve the performance of data mining. Decision support systems help prune the results and give the essential results to carry out management functions.

Sections 4.2 to 4.6 discuss these supporting techniques. Statistical reasoning is addressed in Section 4.2. Machine learning is the subject of Section 4.3. Visualization issues are discussed in Section 4.4. Parallel processing is the subject of Section 4.5. Section 4.6 addresses decision support and Section 4.7 summarizes this chapter.

4.2 STATISTICAL REASONING

Statistical reasoning techniques and methods have been around for several decades. They were the sole means of analyzing data in the past.

Numerous packages are now available to compute averages, sums, and various distributions for several applications. For example, the census bureau uses statistical analysis and methods to analyze the population in a country. More recently, statistical reasoning techniques are playing a major role in data mining. Some argue that the various statistical packages that have been around for quite a while are now being marketed as data mining products. For us this is not an issue, as statistical reasoning plays just one role. For data mining you need the support of various other technologies including organizing and structuring the data. Many of the older statistical packages did not work with large relational databases. However, the packages that are being marketed today are integrated with various databases as illustrated in Figure 4-1.

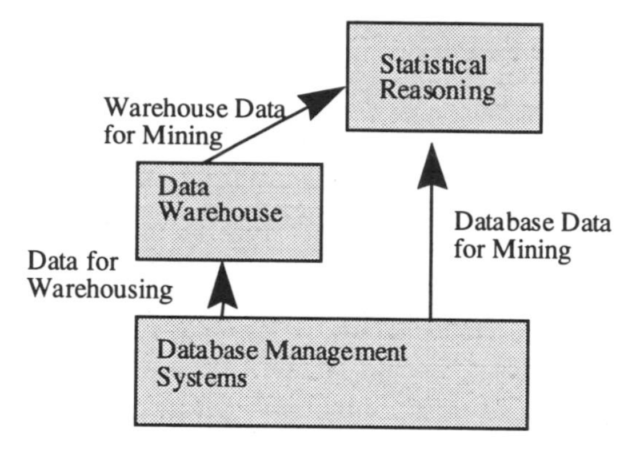

Figure 4-1. Statistical Packages Operating on Databases

As mentioned in [CARB98], the statistical techniques that are being employed for data mining include those based on linear models as well as those based on nonlinear models. Linear regression techniques are employed for prediction. Prediction is a data mining task that predicts variables from available data. For example, one could predict the salary of an employee in five years based on his current performance, his education, and market trends. Linear discriminate analysis techniques are used for classification. Classification is another data mining task where an object is placed in a group based on some classifier. For example, John is placed in a class with others who earn more then 100K. Nonlinear techniques are used to estimate values of new variables based on the data already available and characterized. Estimation is also a data mining task where the various values are estimated based on data that is available. Sampling is another statistical technique used for data analysis. For example, in many cases it will be impossible to analyze all of the data. Therefore, one draws samples such as every Nth row in a relation, forms a sample, and analyzes the sample. There have been some criticisms on using samples for data mining.

In general, one cannot categorically state that linear models are used for classification while nonlinear models are used for estimation. There is no well-defined theory in this area. The point is that statistics plays a major role in data analysis. Even in machine learning, statistics plays a key role. Unless one understands statistics and some operations research, it will be difficult to appreciate machine learning. Because of this, we cannot study data mining without a good knowledge of statistics. This section has discussed statistical reasoning only very briefly. We refer to the numerous texts on this subject should the reader require in-depth knowledge. A useful book is one by DeGroot [DEGR86]. In Mitchell's book on machine learning [MITC97], there is a good introduction to the statistical terms and techniques needed for data mining. These include random variables, probability distribution, standard distribution, and variance.

4.3 MACHINE LEARNING

Machine learning is all about learning rules from the data. Essentially, machine learning techniques are the ones that are used for data mining. So the question is while machine learning has been around for a while, what is new about its connection to data mining? Again the answer is in the data. It is only recently that the various machine learning techniques are being applied to data in databases. These machine learning techniques are soon becoming data mining techniques.

Machine learning is all about making computers learn from experience. As Mitchell describes in his excellent text on machine learning [MITC97], machine learning is about learning from past experiences with respect to some performance measure. For example, in computer games applications, machine learning could be learning to play a game of chess from past experiences which could be games that the machine plays against itself with respect to some performance measure such as winning a certain number of games.

Various techniques have been developed on machine learning. These include concept learning where one learns concepts from several training examples, neural networks, genetic algorithms, decision trees, and inductive logic programming. Each technique is essentially about learning with experience with respect to some performance measure. In Chapter 7 we discuss the various techniques in more detail. We give special consideration to inductive logic programming in Chapter 8 as it is a topic of interest to us. Several theoretical studies have also been conducted on machine learning. These studies attempt to determine the complexity of machine learning techniques [MITC97].

Machine learning researchers have grouped some of the techniques into three categories. One is active learning which deals with interaction and asking questions during learning, the second is learning from prior knowledge, and the third is learning incrementally. There is some overlap between the three methods. Various issues and challenges on machine learning and its relationships to data mining were addressed in a recent workshop on machine learning [DARP98]. There is still a lot of research to be done in this area, especially on integrating machine learning with various data management techniques as shown in Figure 4-2. Such research will significantly improve the whole area of data mining. Some interesting machine learning algorithms are given in [QUIN93].

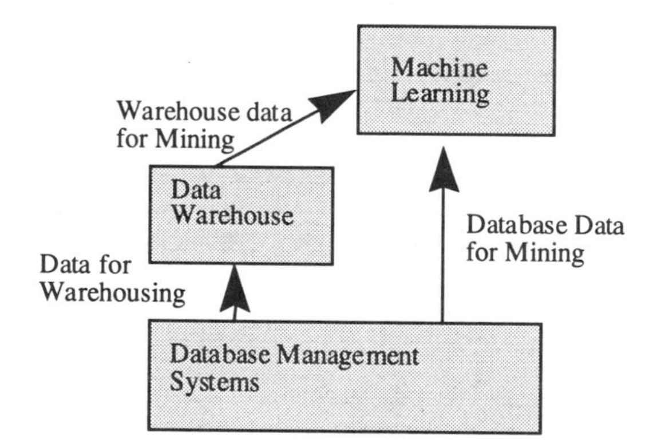

Figure 4-2. Machine Learning and Data Mining

4.4 VISUALIZATION

Visualization technologies graphically display the data in the databases. Much research has been conducted on visualization and the field has advanced a great deal especially with the advent of multimedia computing. For example, the data in the databases could be rows and rows of numerical values. Visualization tools take the data and plot them in some form of a graph. The visualization models could be 2-dimensional, 3-dimensional or even higher. Recently, several visualization tools have been developed to integrate with databases, and workshops are devoted to this topic [VIS95]. An example illustration of integration of a visualization package with a database system is shown in Figure 4-3.

More recently there has been a lot of discussion on using visualization for data mining. There has also been some discussion on using data mining to help the visualization process. However, when considering visualization as a supporting technology, it is the former approach that is getting considerable attention (see, for example, [GRIN95]). As data mining techniques mature, it will be important to integrate them with visualization techniques. Figure 4-4 illustrates interactive data mining. Here, the database management system, visualization tool, and machine learning tool all interact with each other for data mining.

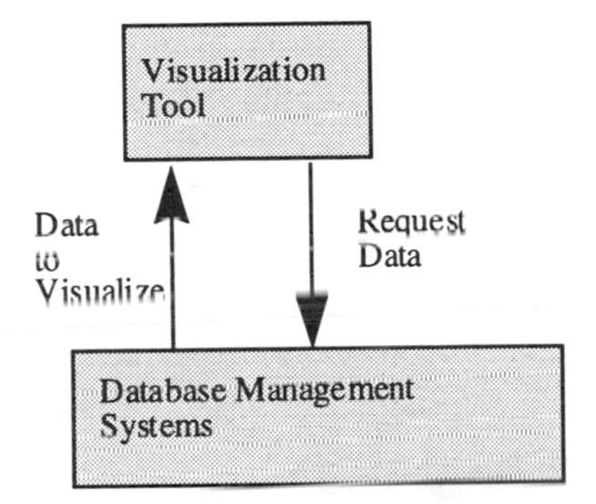

Figure 4-3. Database and Visualization

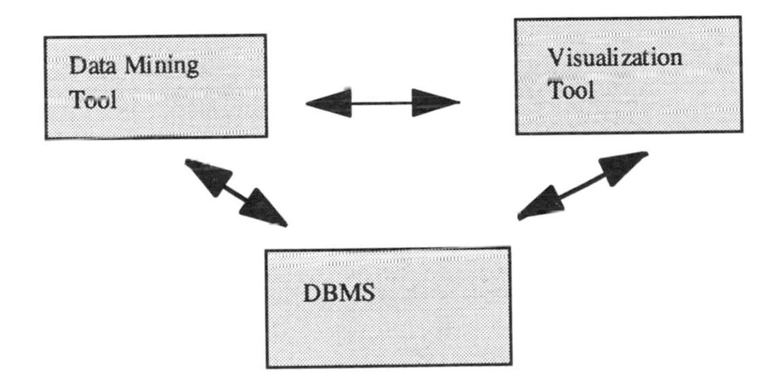

Figure 4-4. Interactive Data Mining

Let us re-examine some of the issues on integrating data mining with visualization. There are four possible approaches here. One is to use visualization techniques to present the results that are obtained from mining the data in the databases. These results may be in the form of clusters or they could specify correlations between the data in the databases. The second approach applies data mining techniques to visualization. The assumption here is that it is easier to apply data mining tools to data in the visual form. Therefore, rather than applying the data mining tools to large and complex databases, one captures some of the essential semantics visually, and then applies the data mining tools. The third approach is to use visualization techniques to complement the data mining techniques. For example, one may use data mining techniques to obtain correlations between data or detect patterns. However, visualization techniques may still be needed to obtain a better understanding of the data in the database. The fourth approach uses visualization techniques to steer the mining process.

In summary, visualization tools help interactive data mining, as was illustrated in Figure 4-4. As illustrated in this figure, visualization tools can be used to visually display the responses from the database system directly so that the visual displays can be used by the data mining tool. On the other hand, the visualization tool can be used to visualize the results of the data mining tool directly. There is little work on integrating data mining and visualization tools. Some preliminary ideas were presented at the 1995 IEEE Databases and Visualization Workshop (see, for example, [VIS95]). However, more progress has been reported in [VIS97]. There is still much work to be done on this topic.

4.5 PARALLEL PROCESSING

Parallel processing is a subject that has been around for a while. The area has developed significantly from single processor systems to multiprocessor systems. Multiprocessor systems could be distributed systems or they could be centralized systems with shared memory multiprocessors or with shared-nothing multiprocessors. There has been a lot of work on using parallel architectures for database processing (see, for example, [IEEE89]). While considerable work was carried out, these systems did not take off commercially until the development of data warehousing. Many of the data warehouses employ parallel processors to speed up query processing.

In a parallel database system, the various operations and functions are executed in parallel. While research on parallel database systems began in the 1970s, it is only recently that we are seeing these systems being used for commercial applications. This is partly due to the explosion of data warehousing and data mining technologies where performance of query algorithms is critical.

Let us consider a query operation which involves a join operation between two relations. If these relations are to be sorted first before the join, then the sorting can be done in parallel. We can take it a step further and execute a single join operation with multiple processors. Note that multiple tuples are involved in a join operation from both relations. Join operations between the tuples may be executed in parallel.

Many of the commercial database system vendors are now marketing parallel database management technology. This is an area we can expect to grow significantly over the next decade. One of the major challenges here is the scalability of various algorithms for functions such as data warehousing and data mining.

Recently parallel processing techniques are being examined for data mining. Many of the data mining techniques are computationally intensive. Appropriate hardware and software are needed to scale the data mining techniques. Database vendors are using parallel processing machines to carry out data mining. The data mining algorithms are parallelized using various parallel processing techniques. This is illustrated in Figure 4-5.

Vendors of workstations are also interested in developing appropriate machines to facilitate data mining. This is an area of active research and development, and corporations such as Silicon Graphics and Thinking Machines have developed products. We can expect to see a lot of progress in this area during the next few years.

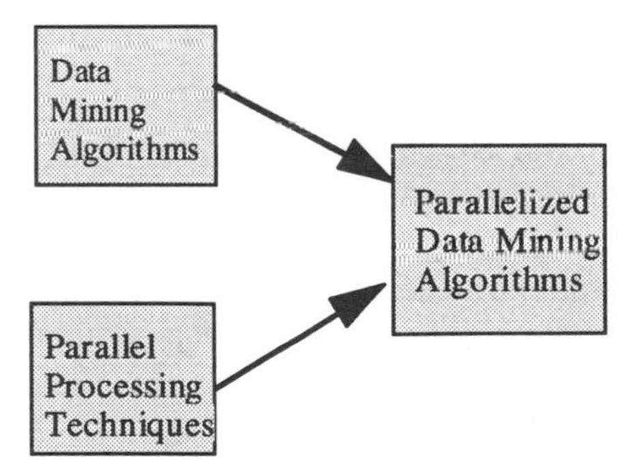

Figure 4-5. Parallel Data Mining

4.6 DECISION SUPPORT

While data mining deals with discovering patterns from the data, machine learning deals with learning from experiences to do predictions as well as analysis. Decision support systems are tools that managers use to make effective decisions. They are based on decision theory. One can consider data mining tools to be special kinds of decision support tools. So are tools based on machine learning, as well as tools for extracting data from data warehouses. Decision support tools belong to a broad category (see, for example, [DECI]).

In general, decision support tools could also be tools that remove unnecessary and irrelevant results obtained from data mining. These pruning tools could also be decision support tools. They could also be tools such as spread sheets, expert systems, hypertext systems, web information management systems, and any other system that helps analysts and managers to effectively manage the large quantities of data and information. More recently there is an area that is emerging called knowledge management. Knowledge management deals with effectively managing an organization's data, information, and knowledge [MORE98a]. This includes storing the information, managing it, as well as developing tools to extract useful information. Some of the knowledge management tools also help in decision support.

In summary, we believe that decision support is a technology that overlaps with data mining, data warehousing, knowledge management, machine learning, statistics, and other technologies that help to manage an organization's knowledge and data. We illustrate this in Figure 4-6. Figure 4-7 illustrates the relationship between data warehousing, database management, mining and decision support.

4.7 SUMMARY

While Chapters 2 and 3 discussed data management and data warehousing, two key supporting technologies for data mining in some detail, this chapter has provided an overview of some of the other supporting technologies. These include statistics, machine learning, visualization, parallel processing, and decision support. Just because we have packed all these supporting technologies into one chapter does not mean that they are any less important. For successful data mining all these technologies have to work together. We have given a lot of attention to data management and warehousing as we have taken a data-centric approach to mining partly because we come from the data management world and partly because unless you have good data, you cannot get good results from mining. Furthermore, devoting a chapter to each of the supporting technologies is beyond the scope of this book. However, we have given various references should the reader require more depth into these supporting technologies.

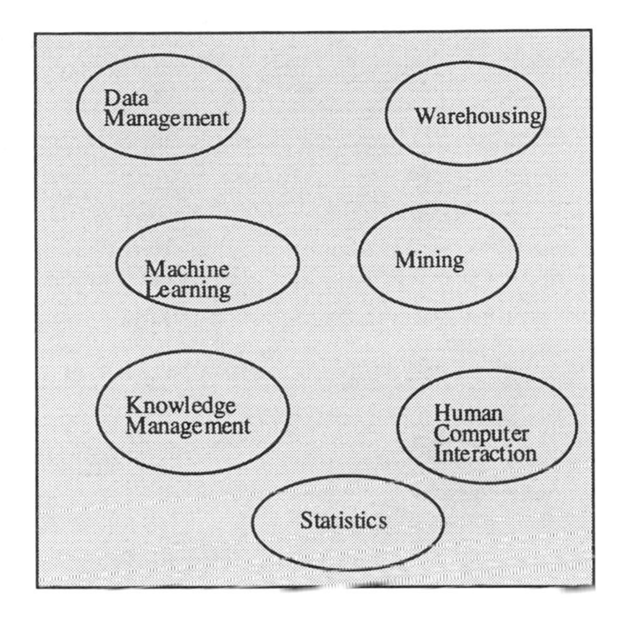

Figure 4-6. Decision Support Technologies

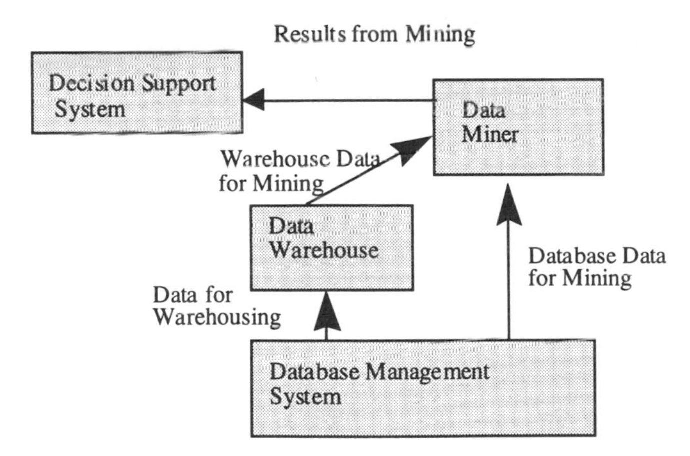

Figure 4-7. Decision Support and Data Mining

We have discussed various data mining technologies and also discussed the relationship of these technologies to data mining. In the next chapter we will discuss architectural aspects for data mining. This will prepare us for the concepts in data mining to be discussed in Part II of this book. It should be noted that other supporting technologies such as agents, human computer interaction, and distributed processing and mass storage exist. Describing all these technologies is beyond the scope of this book. We address some user interface issues in Chapter 6.

CHAPTER 5

ARCHITECTURAL SUPPORT FOR MINING

5.1 OVERVIEW

Previous chapters discussed various supporting technologies for mining. As part of this discussion some architectural aspects were also discussed. For example, one can consider data management and data warehousing to be part of the middleware for data mining. Machine learning and statistics are part of data mining. Then decision support is built on top of data mining. Furthermore, visualization and parallelization are utilized by data mining. We will revisit these issues in this chapter and also address functional as well as system architectures.

Not much work has been done on architectural support for data mining. This is because data mining is still a new area and much of the focus has been on doing mining on relational databases. However, as data mining is applied to other kinds of databases such as object-oriented, multimedia, distributed, heterogeneous, and legacy databases, architectural support will be increasingly important.

In this chapter we will first discuss architectures for integrating with other technologies in Section 5.2. Then in Section 5.3 we will provide an overview of a functional architecture for data mining. Note that in Chapter 6 we will discuss the data mining steps in details. However, in the functional architecture we will only discuss those components that will be part of what we call the data miner. Then in Section 5.4 we will discuss system architectures based on client-server technology for data mining. First we provide an overview of client-server technology and then discuss data mining aspects. The chapter is summarized in Section 5.6.

5.2 TECHNOLOGY INTEGRATION ARCHITECTURE

In Chapters 2, 3, and 4 we discussed various technologies for data mining. One needs architectural support for integrating these technologies. Figure 5-1 shows a pyramid-like structure as to how the various technologies fit with one another. As shown in Figure 5-1, we have communications and system level support at the lowest level. Then we have middleware support. This is followed by database management and data warehousing. Then we have the various data mining technologies. Finally, we have the decision support systems that take the results of data mining and help the users to effectively make decisions. These users could be managers, analysts, programmers, and any other user of information systems.

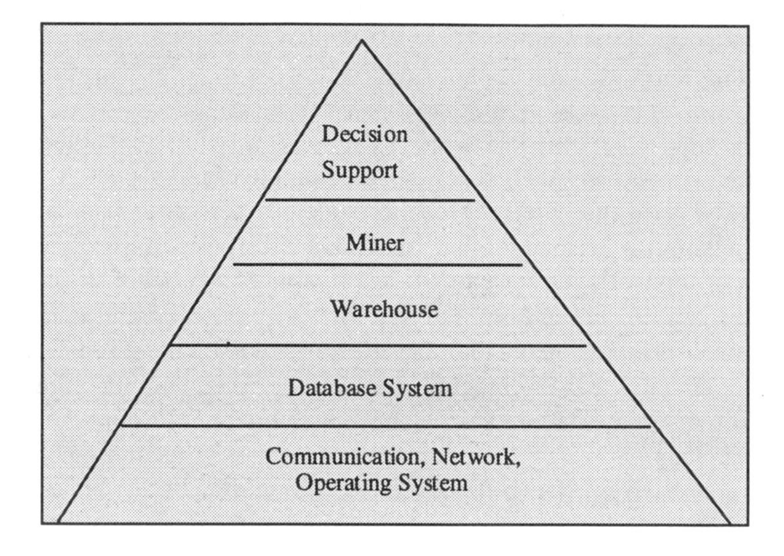

Figure 5-1. Pyramid for Data Mining

When one builds systems, the various technologies involved may not fit the pyramid identically as we have shown. For example, we could skip the warehousing stage and go straight to mining. One of the key issues here is the interfaces between the various systems. At present we do not have any well-defined standard interfaces except some of the standards and interface definition languages emerging from various groups such as the Object Management Group. However, as these technologies mature, one can expect standards to be developed for the interfaces.

Various other diagrams have been illustrated throughout Chapters 1, 2, 3, and 4 as to how the different technologies work together. For example, one possibility is the one shown in Figure 5-2 where multiple databases are integrated through some middleware and subsequently form a data warehouse which is then mined. The data mining component is also integrated into this setting so that the databases are mined directly. Some of these issues will be discussed in the section on system architecture.

Figure 5-3 illustrates a three-dimensional view of data mining technologies. Central to this is the technology for integration. This is the middleware technology such as distributed object management and also web technology for integration and access through the web. On one plane we have all the basic data technologies such as multimedia, relational and object databases, and distributed, heterogeneous and legacy databases. On another plane we have what we call the technologies that do data mining. We have included warehousing as well as machine learning such as inductive logic programming and statistical reasoning here. The third plane has technologies such as parallel processing, visualization, metadata management, and secure access which are important to carry out data mining. The various chapters in this book have discussed the technologies of Figure 5-3 in more detail. The remainder of this chapter discusses functional and system architectures.

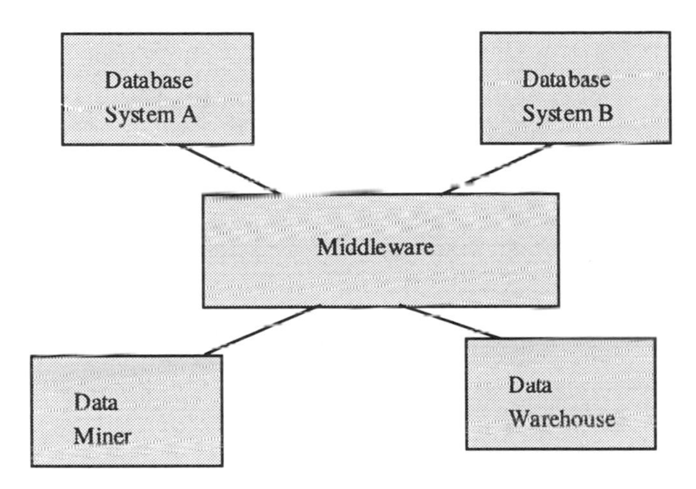

Figure 5-2. Revisiting the Data Mining Architecture

5.3 FUNCTIONAL ARCHITECTURE

The steps to data mining will be elaborated in Chapter 6. These steps also describe the functional components of data mining. These functional components are the subject of this section. Note that in Chapter 2 we discussed the functional components of a database management system. In addition, we illustrated an architecture where the data miner was one of the modules of the DBMS. We called such a DBMS to be a Mining DBMS. There are various ways one can organize a Mining DBMS. An alternative approach is illustrated in Figure 5-4. In this approach we consider data mining to be an extension to the query processor. That is, the query processor modules such as the query optimizer could be extended to handle data mining. This is more of a high level view as illustrated in Figure 5-4. Note that in this diagram we have omitted the transaction manager, as data mining is used mostly for on-line analytical processing.

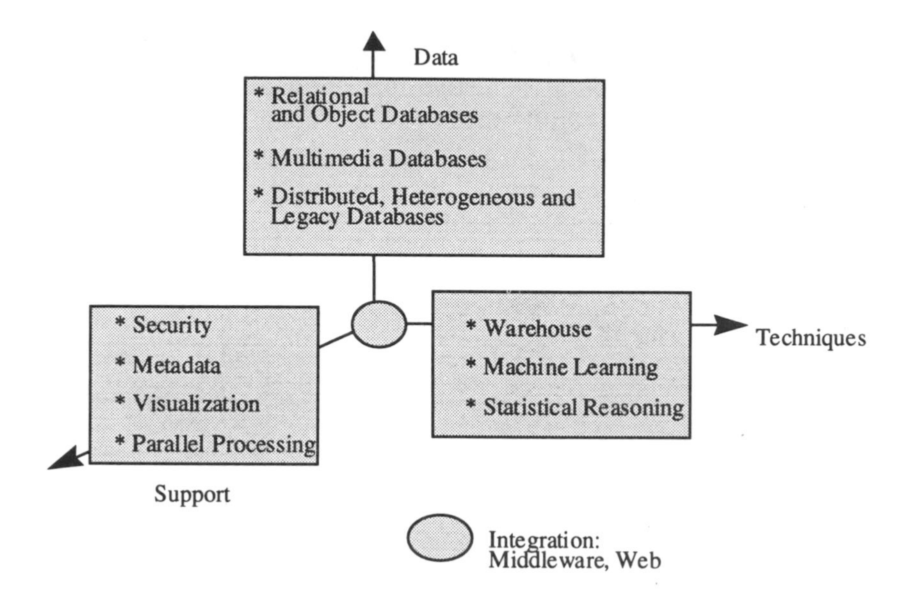

Figure 5-3. Three-Dimensional View

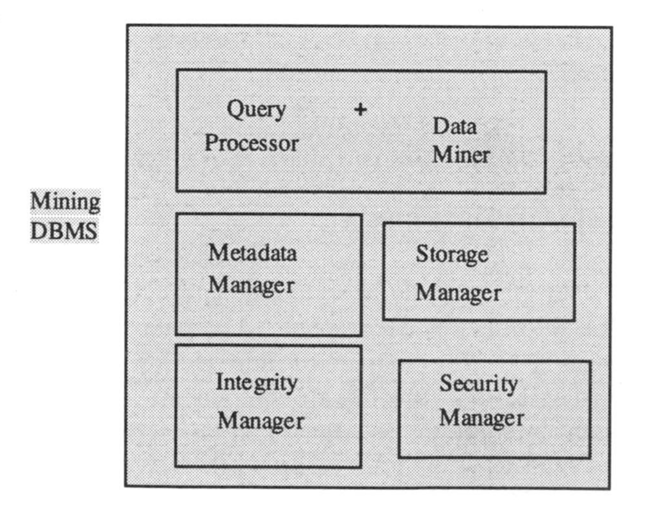

Figure 5-4. Data Mining as Part of Query Processor

The question is what are the components of the data miner? As illustrated in Figure 5-5, a data miner could have the following components: a learning from experience component that uses various training sets and learns various strategies, a data analyzer component which analyzes the data based on what it has learnt, and a results producing component that does classification, clustering, and other tasks such as associations. There is interaction between all three components. For example, the component that produces results then feeds the results back to the training component to see if this component has to be adapted. The training component feeds information to the data analyzer component. The data analyzer component feeds information to the results producing component.

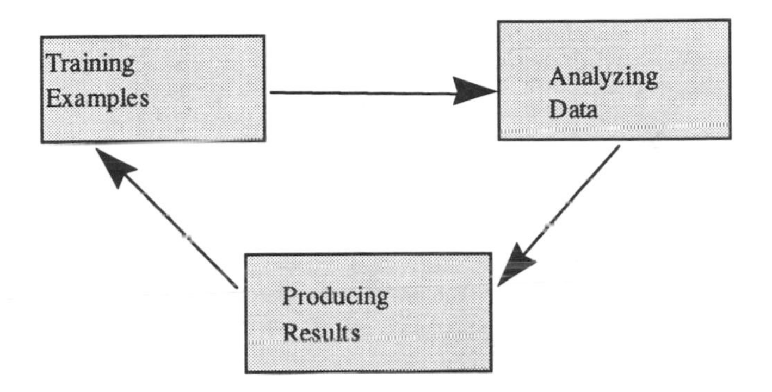

Figure 5-5. Data Mining Functions

Note that we have not included components such as data preprocessor and results pruner into the data mining modules. These components are also needed to complete the entire process. The data preprocessor formats the data. In a way the data warehouse may do this function. The results pruning component may extract only the useful information. This could be carried out by a decision support system. All of these steps will be integrated into the data mining process and discussed in Chapter 6.

5.4 SYSTEM ARCHITECTURE

5.4.1 Overview

Some of the architectures we discussed in Chapters 1, 2, and 3, as well as that in Figure 5-2 can be regarded to be a system architecture for data mining. A system architecture consists of components such as the middleware and other system components such as the database management system and the data warehousing system for data mining.

The middleware that we have illustrated in Figure 5-2 could be based on various technologies. One popular middleware system is that based on client-server architectures. In fact, many of the database systems are based on client-server architectures. Middleware also includes de facto standards such as Microsoft's Open Database Connectivity (ODBC) or distributed object-based systems.

In [THUR97] we provided a detailed discussion of client-server technologies. In particular we discussed the client-server paradigm as well as provided an overview of ODBC and distributed object management systems such as the Object Management Group's (OMG) Common Object Request Broker Architecture (CORBA). In this section we discuss data mining with respect to client-server paradigm. First in Section 5.4.2 we summarize client-server technologies. Then in Section 5.4.3 we discuss data mining and client-server technologies.

5.4.2 Client-Server Technology

Major database system vendors have migrated to an architecture called the client-server architecture. With this approach, multiple clients access the various database servers through some network. A high level view of client-server communication is illustrated in Figure 5-6. The ultimate goal is for multi-vendor clients to communicate with multi-vendor servers in a transparent manner.

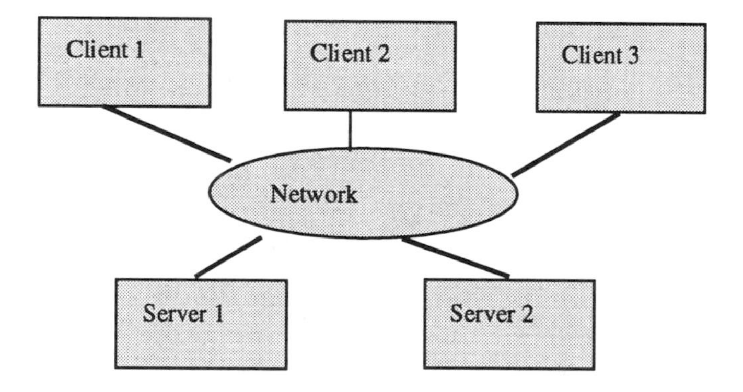

Figure 5-6. Client-Server Architecture-Based Interoperability

In order to facilitate the communication between multiple clients and servers, various standards are being proposed. One example is the International Standards Organization's (ISO) Remote Database Access (RDA) standard. This standard provides a generic interface for communication between a client and a server. Microsoft ODBC is also becoming increasingly popular for clients to communicate with servers. OMG's CORBA provides specifications for client-server communication based on object technology. Here, one possibility is to encapsulate the database servers as objects and the clients to issue appropriate requests and access the servers through an Object Request Broker (ORB). Other standards include IBM's DRDA (Distributed Relational Database Access) and the SQL Access Group's (now part of the Open Group) Call Level Interface (CLI). Quite a few books have been published recently on client-server computing and data management. Two good references are [ORFA94] and [ORFA96]. We have also discussed some of these issues in detail in [THUR97].

A middleware system that is becoming increasingly popular to connect heterogeneous systems is OMG's CORBA. As stated in [OMG95], there are three major components to CORBA. One is the object model which essentially includes most of the constructs discussed in Chapter 2, the second is the Object Request Broker (ORB) through which clients and servers communicate with each other, and the third is the Interface Definition Language (IDL) which specifies the interfaces for client server communication. Figure 5-7 illustrates client server communication through an ORB. Here, the clients and servers are encapsulated as objects. The two objects then communicate with each other. Communication is through the ORB. Furthermore, the interfaces must conform to IDL.

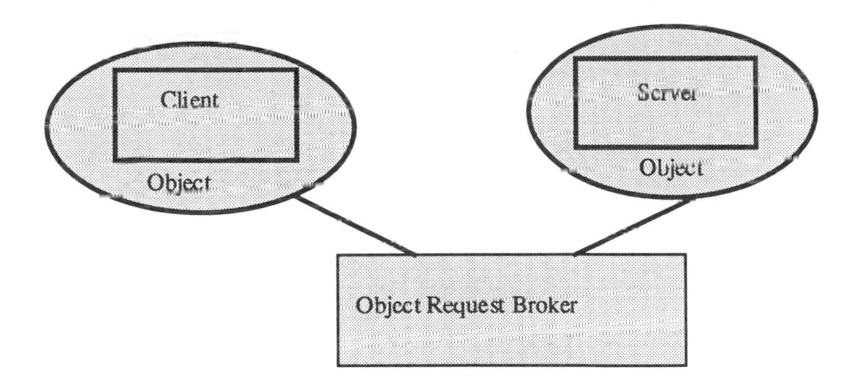

Figure 5-7. Interoperability through ORB

Since data mining is the subject of this book, in the next section we will illustrate how client-server architectures could be utilized for data mining.

5.4.3 Relationship to Mining

Consider the architecture of Figure 5-8. In this example, the data miner could be used as a server, the database management system could be another server, while the data warehouse could be a third sever. The client issues requests to the database system, warehouse, and the miner as illustrated in this figure.

One could also use an ORB for data mining. In this case the data miner is encapsulated as an object. The database system and warehouse are also objects. This is illustrated in Figure 5-9. The challenge here is to define IDLs for the various objects.

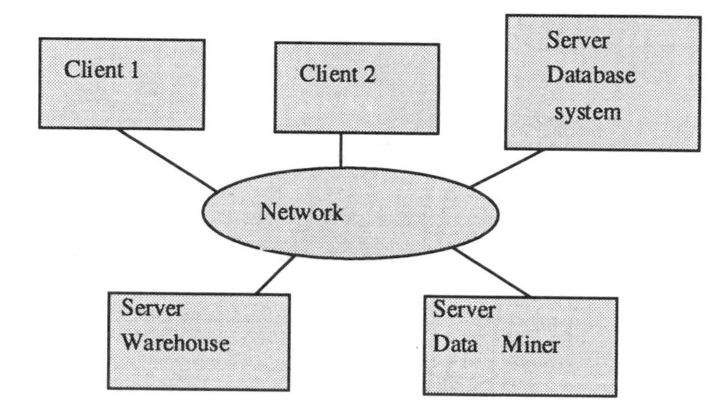

Figure 5-8. Client-server-based Data Mining

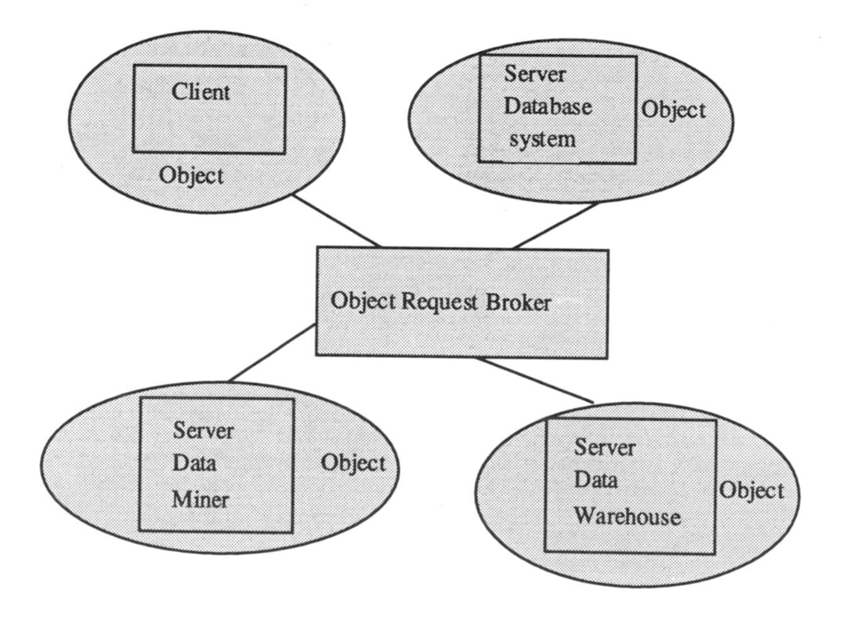

Figure 5-9. Data Mining through ORB

Note that client-server technology does not develop algorithms for data management, warehousing, or mining. This means that the algorithms are still needed for mining, warehousing, and database management. What client-server technology and, in particular, distributed object management technology such as CORBA do for you is facilitate interoperation between the different components. For example, the data miner, database system, and warehouse communicate with each other and with the clients through the ORB.

Note that the three-tier architecture is becoming very popular (see the discussion in [THUR97]). In this architecture, the client is a thin client and does minimum processing, the server does the database management functions, and the middle tier carries out various business processing functions. In the case of data mining, one could also utilize a three-tier architecture where the data miner is placed in the middle tier as illustrated in Figure 5-10. The data miner could be developed as a collection of components. These components could be based on object technology. By developing data mining modules as a collection of components, one could develop generic tools and then customize them for specialized applications.

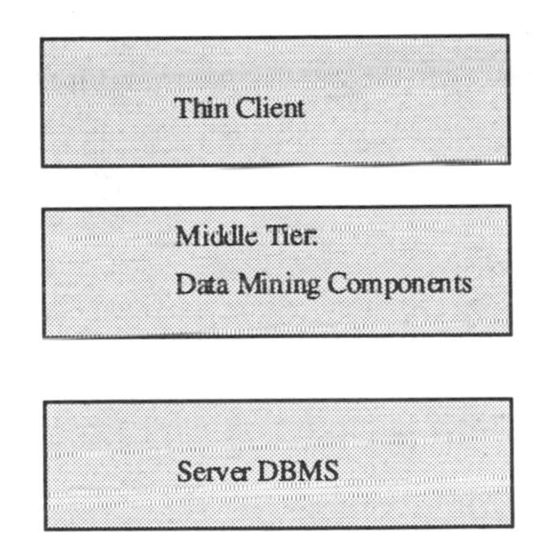

Figure 5-10. Three-Tier Architecture for Data Miner

Another advantage of developing a data mining system as a collection of components is that one could purchase components from different vendors and assemble them together to form a system. Furthermore, components can be reused. We will elaborate on the various data mining modules in Chapter 6. For now let us assume that the modules are the data source integrator, the data miner, the results pruner, and the report generator. Then each of these modules can be encapsulated as objects and one could use ORBs to integrate these different objects. As a result, one can use a plug-and-play approach to developing data mining tools. Figure 5-11 illustrates the encapsulation of the various data mining modules as objects. One could also decompose the data miner into multiple modules and encapsulate these modules as objects. For example, consider the modules of the data miner illustrated in Figure 5-5. These modules are part of the data miner module and could themselves be encapsulated as objects and integrated through an ORB.

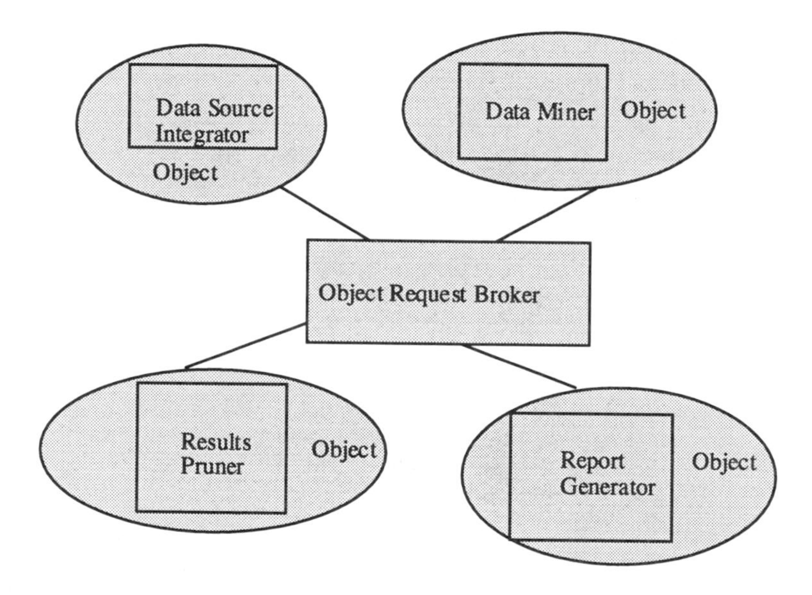

Figure 5-11. Encapsulating Data Mining Modules as Objects

5.5 SUMMARY

This chapter has discussed various architectural aspects for data mining. We started with a discussion of architectures for technology integration. Then we discussed functional architectures for data mining. Finally, we provided an overview of client-server technology for data mining. In particular, we discussed the ORB technology and the three-tier architecture for client-server systems and showed such technologies and architectures could be utilized for data mining systems. We believe that objects and components will play a major role in developing data mining tools in the future. With this approach, one could purchase different components and assemble them to develop data mining tools for special applications.

It should be noted that since data mining is still a new field, there is little work on architectures for data mining. The discussion in this chapter is still preliminary. Nevertheless, functional as well as system architectures will be an important aspect of data mining, and we can expect to see developments in this area. Furthermore, one can expect the architectural developments in database management systems to influence the developments in data mining.

Conclusion to Part I

The data mining technologies discussed in Part I of this book constitute the technologies of Layer 1 of the data mining framework. These technologies are database management, data warehousing, statistical reasoning, machine learning, visualization, parallel processing, and decision support. In addition, architectural support for data mining is also a supporting technology and is part of Layer 1. Data mining is built on these supporting technologies.

Now that we have provided some information on the background for data mining, we are ready to explain what data mining is all about. In particular, the process of data mining, the outcomes of data mining, various data mining techniques, and the tools that have been developed will be discussed next. The essential points of data mining are the subject of Part II of this book.

Part II

Data Mining Techniques and Tools

Introduction to Part II

Part II, consisting of four chapters, describes techniques and tools for data mining. In Chapter 6 we start with a discussion of the steps for data mining. These steps include preparing the data, selecting the tools, carrying out the mining, post processing of the results, and taking actions and measuring success. Then in Chapter 7 we discuss outcomes and techniques for data mining. Data mining outcomes include classification, clustering, estimation, prediction, and associations/correlations. The outcomes are what can be expected as a result of data mining. Data mining techniques are the algorithms employed for doing the mining. These include neural networks, decision trees, and genetic algorithms. We pay special attention to a particular data mining technique based on inductive logic programming. This is mainly due to our interest in this topic. Inductive logic programming and data mining is the subject of Chapter 8. Then in Chapter 9 we provide an overview of the various prototypes and commercial tools in data mining. It should be noted that due to the rapid advances, the information on prototypes and products could soon be outdated. Therefore we encourage the reader to keep up with product information.

While Part I provided background information for data mining, Part II describes what data mining is all about. This will prepare us for the discussions in Part III that deal with special topics in data mining such as privacy and security issues, distributed data mining, multimedia data mining, and web mining.

CHAPTER 6

THE PROCESS OF DATA MINING

6.1 OVERVIEW

Now that we have an understanding of what the data mining technologies are and how they contribute to data mining, let us next discuss what data mining is all about and how we go about doing mining. Remember that data mining is the process of posing queries and extracting useful information, patterns and trends previously unknown from large quantities of data stored possibly in databases. That is, not only do we want to get patterns, and trends, these patterns and trends must be useful, else we can get irrelevant data that could turn out to be harmful or cause problems with the actions taken. For example, if an agency finds incorrectly that its employee has carried out fraudulent acts and then starts to investigate his behavior, and if this is known to the employee, then it could damage him. This is called a false positive. However, we also do not want results that are false negatives. That is, we do not want the data miner to return a result that the employee was well behaved when he is a fraud. So data mining has serious implications. This is why it is critical that we have good data to mine and we know the limitations of the data mining techniques.

We would like to stress to managers and project leaders not to rush into data mining. Data mining is not the answer to all problems and sometimes it has been over emphasized. It is expensive to carry out the entire mining process and therefore has to be thought out clearly. In reality, the mining part is only a small step toward the entire process. One needs to ask various questions such as is there a need for mining? Do you have the right data in the right format? Do you have the right tools? More importantly, do you have the people to do the work? Do you have sufficient funds allocated to the project? All these questions have to be answered before you embark on a data mining project. Otherwise you can be extremely disappointed with the results.

This chapter discusses data mining from start to finish without going into the technical details such as algorithms, approaches, and outcomes. These details are given in the next chapter. In Section 6.2, we describe various example applications that would benefit from data mining. These are examples we have obtained from the various discussions we have had, articles read, and using common sense. This would give the reader a good flavor as to what data mining is all about. In Section 6.3, we discuss why we want to do data mining. For example, why is data mining such a buzz word now? What is it about the world that has changed that makes data mining so useful now and not twenty years ago when we still had computers? Then in Section 6.4, we discuss the steps to data mining. These steps include identifying and preparing the data, determining which data to mine, preparing the data to mine, carrying out the mining, pruning the results, taking actions, evaluating the actions, and determining when to do mining next. Note that data mining is not a one time activity. An organization has to continually do mining as the data may have changed, the actions may not be beneficial, or the tools may have improved. Then in Section 6.5, we discuss some of the limitations and challenges to data mining. These include incorrect data, incomplete data, insufficient resources such as man power, and inadequate tools. Some user interface aspects are discussed in Section 6.6. Finally, we summarize the chapter in Section 6.7.

6.2 EXAMPLES

In this section we give numerous examples to illustrate how mining can be used. Some of these examples have been obtained from various papers and proceedings (see, for example, [GRUP98]) and some others from discussions. While much of the work in data mining is being done to support marketing and sales, it is also useful in other areas. Here are some examples.

- A supermarket store analyzes the purchases made by various people and arranges the items on the shelves in such a way to improve sales.

- A credit bureau analyses the credit history of various people and determines who are at risk and who are not.

- An investigation agency analyzes the behavior patterns of people and determines who can be potential threats to protected information.

- A pharmacy determines which physicians are likely to buy their products by analyzing the prescription patterns of physicians.

- An insurance company determines the patients who might be potentially expensive by analyzing various patient records.

- An automobile sales company analyzes the buying patterns of people living in various locations and sends them brochures of cars that customers are likely to buy.
- An employment agency analyzes the various employment history of employees and sends them information of potentially lucrative jobs.
- An adversary uses data mining tools and gets access to unclassified databases and by some way deduces information potentially classified.
- An educational institution analyzes student records and determines who are likely to attend their institution and sends them promotional brochures.
- A nuclear weapons plant analyzes audit records of history information and determines that there could be a potential nuclear disaster if certain precautions are not taken.
- A command and control agency analyzes the behavior patterns of adversaries and determines the weapons that the adversary has.
- A marketing organization analyzes the buying patterns of various people and estimates the number of children they have and their income so that potentially useful marketing information can be sent.
- By analyzing the travel patterns of various groups of people an investigation agency determines the associations between the various groups.
- By analyzing patient history and current medical conditions, physicians not only diagnose the medical conditions but also predict potential problems that could occur.
- The income tax revenue office examines the tax returns of various groups of people and finds abnormal patterns and trends.
- An investigation agency analyzes the records of criminals and determines who are likely to commit terrorism and mass murders.

Note that we have selected various examples from all types of applications including from financial, intelligence, and medical to carry out various activities such as marketing, diagnosis, correlations and fault detection. These application areas are illustrated in Figure 6-1. All these examples show that a great amount of data analysis is needed to come up with the results and conclusions. This type of data analysis is what is usually referred to as mining. Note that mining is not always used in a positive manner to better the human society. In many cases it could be quite dangerous such as compromising the privacy of individuals. In this book we have provided information not only on the positive aspects but also on the negative aspects. In addition, we also discuss the difficulties and challenges to mining.

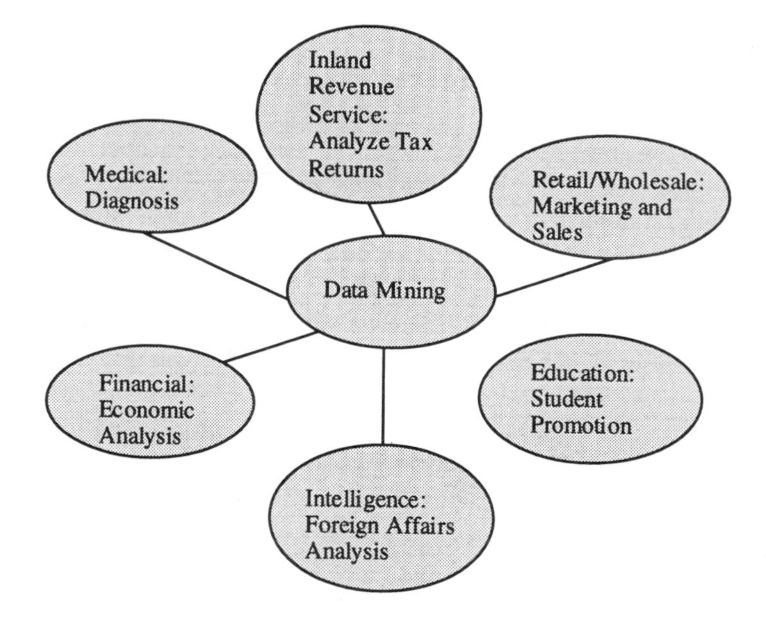

Figure 6-1. Some Data Mining Application Areas

6.3 WHY DATA MINING?

So now that we have seen various applications that may need data mining, let us discuss why is it that we are talking about data mining now? We know that many of these problems have existed for several years. Furthermore, data has been around for centuries. The answer to this is that we are using new tools and techniques to solve problems in a new way. These problems have existed and people have worked on handling them for years. For example, data analysis has been carried out for years in different ways, but it is only now that we call it data mining with improved methods and techniques (see Figure 6-2).

Although data has been around, it has been on paper and in many cases not even on paper but in the minds of people. Typically clerks spend years recording the data, and human analysts go through the data to detect various patterns. Then the whole area of statistics started and that gave a new way to analyze the data. However, organizing the data was still a big problem. Then with computers, and especially databases, we started storing the data in computerized files and databases. This was the first big step toward data mining. Then came the area of artificial intelligence with new and improved searching and learning techniques. What has contributed mostly to data mining is the improved way to store and retrieve the data, and that is essentially database management systems technology.

More recently, techniques and tools are being developed to focus on improving methods to capture the data and knowledge of organizations. This is going to be even better for data mining. There are still lots and lots of data out there that have not been captured. Furthermore, even if it is captured, one does not know of the existence of the data. So, knowledge management techniques would hopefully improve these deficiencies [MORE98a].

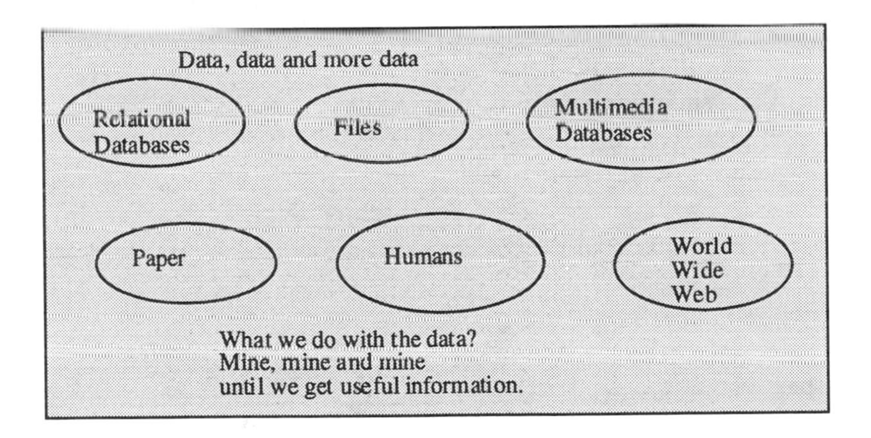

Figure 6-2. Why Data Mining?

So now we have large quantities of data computerized. The data could be in files, relational databases, multimedia databases, and even on the world wide web. We have very sophisticated statistical analysis packages. Tools have been developed for machine learning. Parallel computing technology is getting mature for improving performance. Visualization techniques improve the understanding of the data. Decision support tools are also getting mature. So what better way is there than integrating these various developments to provide improved capabilities for analyzing data and predicting trends? Data mining has become a reality now, and therefore, we are beginning to get ready for data mining.

6.4 STEPS TO DATA MINING

We have given various examples of data mining and established a need for data mining. So, what are the steps to mining? Where do we start and where do we end? Various texts have discussed the steps to data mining and we have found them quite useful (see, for example, [BERR97]). Based on what we have read and from our experiences, the data mining steps, some of which are as illustrated in Figure 6-3, are the following.

- Identifying the data
- Getting the data ready
- Mining the data
- Getting useful results
- Identifying actions
- Implementing the actions
- Evaluating the benefits
- Determining what to do next
- Carrying out the next cycle

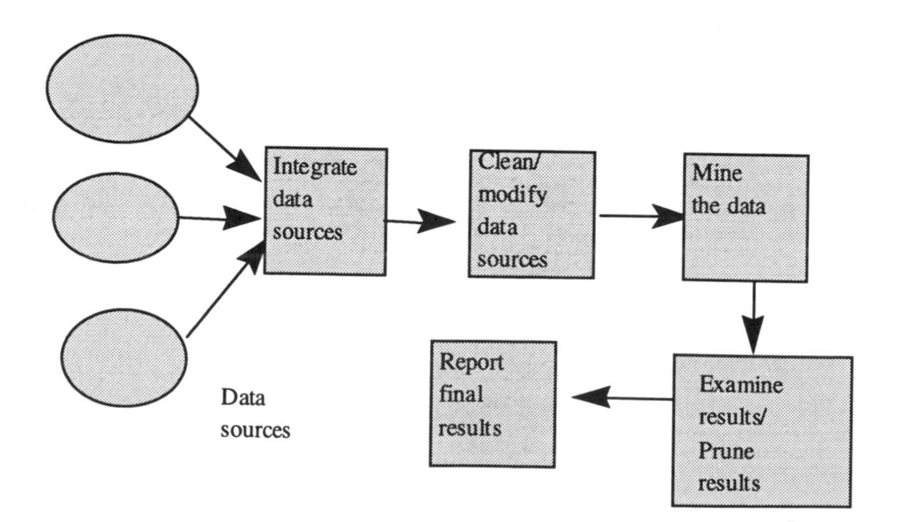

Figure 6-3. Steps to Mining

We will elaborate on each step. First we need to identify the data. As mentioned in the previous section, data could be all over the world not just in the enterprise. Data could be distributed. It could be on paper and even in people's heads. We need to figure out what data we need and where we can find it, then go and get the data.

Once we have the data, we need to prepare the data. There is a lot of effort in this. We may have to put it in databases in the right format. Even worse, we may have to build a data warehouse or get a database management system. This is by no means trivial and people under estimate this step. We have repeatedly heard people say that this is one of the most difficult tasks in mining.

So now we have the data in the right format. We also need to clean the data, scrub unnecessary items, and get only the data essential for mining. This is also not trivial. We now have the data we want to mine and we have seen how to go about mining. Next, what outcomes do we want? It is good to have some idea. Do we want the tools to find interesting patterns without letting it know something about what we want? Then how do we go about getting what we want? Are there tools available? Do we build the tool? This is very time consuming also.

So now we have determined the data and the tool. Next, we start the tool to operate on the data. This is sort of the easy part. The tool could produce lots and lots of data, which may seem like a foreign language to many. So now what do we do with the patterns? Do we have an application specialist analyze the patterns? Are there analysts who can figure out what the data is all about? Do we have tools to analyze the results and get useful patterns? That is, we need to figure out how to effectively prune the results and get only the useful results. The pruning process is illustrated in Figure 6-4.

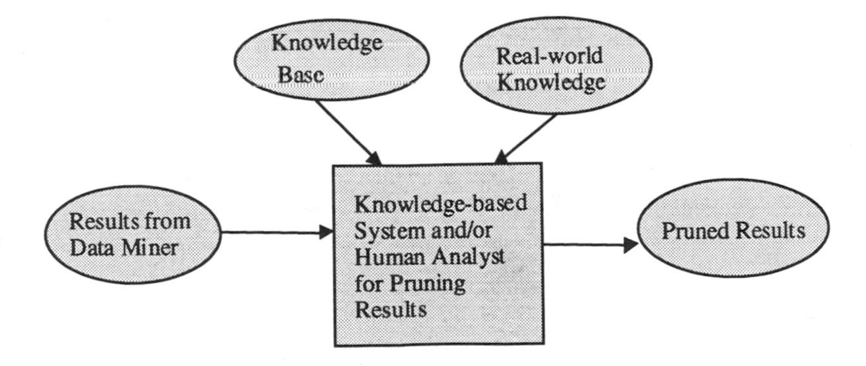

Figure 6-4. Pruning the Results

So now we think we have useful results. We now need to examine the results and identify actions that can be taken. For example, in a supermarket store, by analyzing the various purchases, we decide that we need to put milk and moisturizing cream together. So now we identify the various actions we think will be beneficial and then discuss procedures to implement the actions.

We now have the actions implemented. We wait to see the results. The results could be immediate or take a very long time. Once we are in a position to determine the benefits and costs of our actions, we then reevaluate the whole procedure. By then the data may have changed. New tools may be available. We may have to do things differently. So we plan for the next mining cycle and determine how to go about doing it.

Note that the above discussion does not bring the human element into the process. Humans play a major part. First of all, we do need management buy in. Therefore we must be very careful not to oversell the project. Be realistic as to what mining can and cannot do for us. Once management is convinced, we need to determine whether we have the man power. Do we train people or contract it out? Also we need to discuss these issues with our customer. Another question is, what do we do about the tool developers? Are they from the outside or from inside of our corporation? The customer, contractor, and tool developer have to work very closely to make mining a success.

If the project has failed, then do not try to point the finger at one person or a group of people. Remember this is still a new technology, so the likelihood of success may not be high. Learn from your experiences. Talk to people who have had similar experiences. See what can be done differently. In many cases it might be good to start a small pilot project or prototype effort before going into full scale mining.

6.5 CHALLENGES

So now we know the applications that need mining, why we are mining, and the steps involved in mining. Next, what are the difficulties both technical and otherwise? The non-technical difficulties include not enough management support and resources, such as trained individuals and low budgets. The technical challenges are many. We discuss some of them.

We have stressed that getting the right data in the right format is critical. But there are lots of problems here. First of all, the data may not be accurate. What do we do then? How do we track down the source, as data may have passed many levels? This is one of the major challenges. Then the data may be incomplete. There may be lots of missing values. So how do we fill in the blanks? The data may be uncertain. That is, one may not be certain how accurate the data is. Here again, do we track down the source? So, missing, inaccurate and uncertain data are major challenges.

Next, do we have the right tools? If not, what do we do? Do we get tools and adapt them or develop them from scratch? This is a big problem since the tools are still not mature. Another challenge is developing adaptive techniques. That is, many of the tools only do one type of data mining like classification or clustering. Can we develop tools that can adapt to the situation and carry out a particular type of mining? Can a tool use multiple mining techniques and handle different outcomes? There is research in this area, but we are a long way from robust commercial tools.

We have named a few of the challenges and some of them are illustrated in Figure 6-5. The good news is that there is now a lot of research in data mining and now initiatives are being formed. So, as time goes by, we feel that we will be getting good answers to many of the questions we have posed.

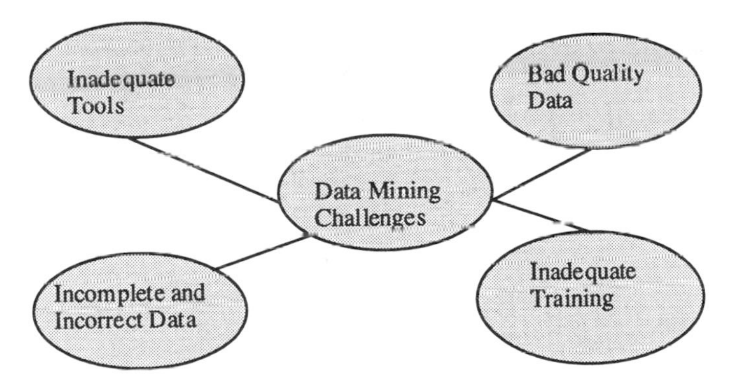

Figure 6-5. Some Data Mining Challenges

6.6 USER INTERFACE ASPECTS

As in any kind of system, having a good user interface is critical to mining. Note that some of the early database management systems had very primitive user interfaces. Therefore, users had to spend a great deal of time writing SQL queries and applications programs. After much work, current database systems have excellent user interface tools. These include tools for generating queries, applications programs, as well as reports. Various multi-modal interfaces are also being provided for database management.

User interface support for current data mining systems is fairly primitive. As mentioned in Chapter 4, visualization tools are being developed to help with data mining, but tools for generating queries, application programs to carry out data mining, and reports are not sophisticated. To make data mining a success we need better user interface tools. Computer scientists and technologists are not the only ones who should be involved in developing such tools. Interactions between technologists, scientists, psychologists, and human computer specialists are necessary to develop better tools. Figure 6-6 illustrates an example user interface for data mining. The interface has buttons not only for generating data mining queries, applications for data mining, and reports, but also for selecting the outcomes desired, approaches to be followed, and the techniques to be utilized.

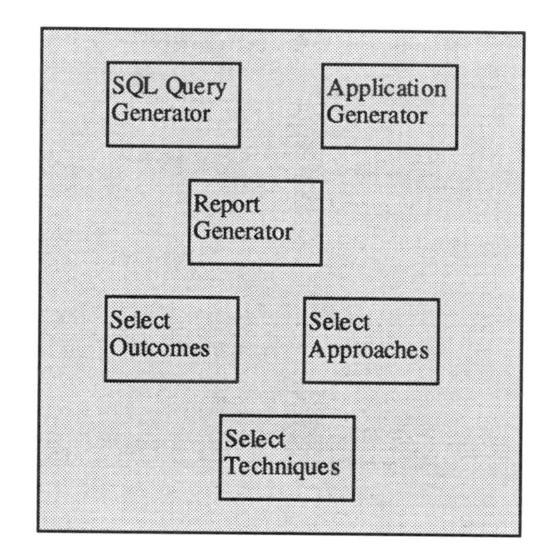

Figure 6-6. Example User Interface for Mining

6.7 SUMMARY

This chapter has discussed the data mining process from start to finish. We started with a discussion of the various examples so that the reader can have a feel for what data mining can do. These examples were selected from different domains to illustrate different outcomes. These included deviation analyses, correlations and associations, and classifications. Then we discussed the reasons for data mining. Note that the applications have existed for a while and even many of the data mining techniques have been around. So why are we paying so much attention to data mining? The answer lies in the fact that we have lots and lots of data available now. This data is also organized and formatted. We have database systems and warehouses to manage the data. Next, we discussed the steps to data mining. These included getting the data ready, carrying out mining, pruning the results, identifying actionable items, carrying out the actions, evaluating outcomes, and determining the next cycle. Next, we discussed the challenges for data mining. These included having incomplete and inaccurate data, insufficient tools and resources, no management commitment, and continually changing data. Finally, we provided some information on user interface aspects.

Data mining has come a long way over the last few years. It emerged as a technology area in the early 1990s. Today we have quite a few data mining tools and prototypes. There is much progress on integrating data management with statistical reasoning and machine learning. The major challenge now is getting useful results from mining. That is, how do you look for what is called "the nuggets?" This is an area that still needs a lot of work, and we need to start some research programs on this topic. Various other books have also addressed the process of data mining. These include [BERR97] and [ADRI96].

CHAPTER 7

DATA MINING OUTCOMES, APPROACHES, AND TECHNIQUES

7.1 OVERVIEW

The last chapter described the steps involved in data mining. These include preparing the data, determining what outcomes to expect, selecting data mining tools, doing the mining, pruning the results, determining the actions to be taken, and evaluating the benefits. This chapter focuses on concepts in data mining. In particular, what possible outcomes can one expect, what the are approaches or methodologies used, and what are the data mining techniques used? Various data mining and machine learning text books, such as the ones by Mitchell [MITC97], Berry and Linoff [BERR97], and Adriaans and Zantinge [ADRI96] have focused mainly on the topics discussed in this chapter. In particular, Berry and Linoff [BERR97] have given an excellent discussion on the outcomes, approaches and techniques for data mining.[13] Therefore, we will only discuss the essential points in this chapter. For further reading we refer the reader to the references we have given.

The outcomes of data mining are also referred to as the data mining tasks or types. These are the results that one can expect to see as a result of data mining. We discussed some of the tasks in Chapter 1. The data mining outcomes include classification, clustering, prediction, estimation, and affinity grouping. It should be noted that there is no standard terminology. Therefore, different papers and texts have used different terms sometimes to mean the same concept.

The approaches to data mining, also referred to as methodologies, are either top-down, bottom-up, or hybrid. In addition, the methods could also be directed or undirected. Directed techniques are also sometimes called supervised learning, while undirected techniques are called unsupervised learning.

Data mining techniques are the algorithms employed to carry out data mining. There has been some confusion between techniques and outcomes. For example, a collection of data mining techniques is used

---

[13] In fact, Berry and Linoff call these terms data mining tasks, methodologies, and techniques in [BERR97]. As we have mentioned, there is no standard terminology for data mining. We hope that the data mining community will eventually standardize various terms.

for market basket analysis. These techniques are now known as market basket analysis techniques. However, market basket analysis is also an application. This is all about determining which items are purchased together in a supermarket. Therefore, there is some confusion within the community as to what these terms mean. We expect progress to be made with respect to terminology as the technology matures and standards are developed.

In general, to carry out data mining for a specific application, first we have to decide on the type of outcome expected from the process. Then we have to determine the techniques to be employed to get the expected outcome. Finally, we have to determine whether to steer the process in a top-down, bottom-up or hybrid fashion. Figure 7-1 illustrates these steps.

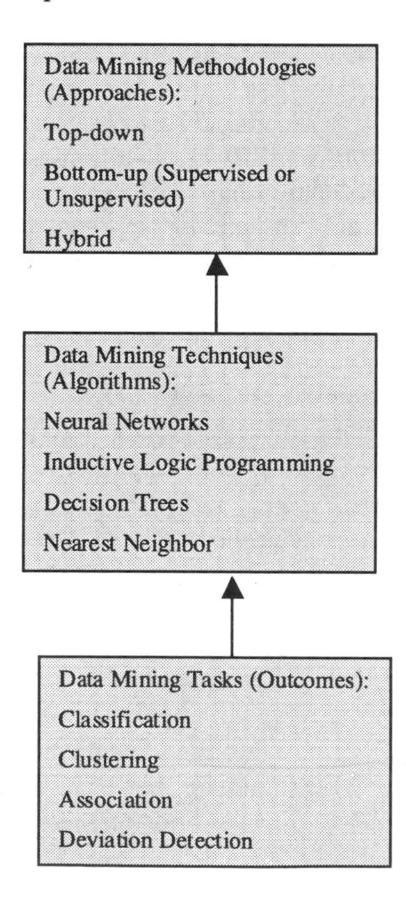

Figure 7-1. Data Mining Tasks, Techniques, and Methodologies

This chapter is devoted to the outcomes, approaches, and techniques for data mining. The outcomes are discussed in Section 7.2. Approaches are the subject of Section 7.3. Data mining techniques are addressed in Section 7.4. The chapter is summarized in Section 7.5.

7.2 OUTCOMES OF DATA MINING

The outcomes of data mining are also referred to as data mining tasks or types. We discuss some of them here. In a task called classification, the tool examines the features of a new entity, examines a predfined set of classes and classifies the entity as belonging to a particular class if common features are extracted. For example, a class of mammals could have attributes that describe a mammal. If a living entity has to be classified and satisfies the properties for a mammal, then it can be classified to be a mammal. Classification is carried out by developing training sets with preclassified examples and then building a model that fits the description of the classes. Then this model is applied to the data not yet classified and results are obtained. In summary, with classification, a group of entities is partitioned based on a predefined value of some attribute. Examples of conditions used to classify a group of people include "age is more than 50," "salary is more than 100K," "is married," and "owns a car costing more than 20K." Note that the classes can be divided further depending on the value of some other attribute. This way, a class hierarchy is formed as in the case of the object class hierarchy described in Chapter 2.

Estimation and prediction are two other data mining tasks. In the case of estimation, based on the spending patterns of a person and his age, one can estimate his salary or the number of children he has. Prediction tasks predict the future behavior of some value. For example, based on the education of a person, his current job, and the trends in the industry, one can predict that his salary will be a certain amount by year 2005.

One task that is extremely useful is affinity grouping. This is also sometimes referred to as making associations and correlations. Essentially this determines the items that go together. Who are the people that travel together? What are the items that are purchased together? While prediction is some future value and estimation is an estimated value, affinity grouping makes associations between current values.

Clustering is a data mining task that is often confused with classification. While classification classifies an entity based on some predefined values of attributes, clustering groups similar records not based on some predefined values. That is, when you classify a group of people, you essentially have predefined classes based on values of some attributes. In the case of clustering you do not have these predefined classes. Instead you form clusters by analyzing the data. For example, suppose you want to find something interesting about a group of people in a community. You do not have any predefined classes. You then analyze their spending patterns on automobiles. Then you form the following clusters based on the analysis: Group X prefers Volvos, Group Y prefers Saabs, and Group Z prefers Mercedes. Once the clusters are obtained, then each cluster can be examined and mined further for other outcomes such as estimation and classification.

Other data mining tasks include deviation analysis and anomaly detection. For example, John usually goes shopping after he goes to the bank, but last week he went to church after shopping. Anomaly detection is a form of deviation detection and is used for applications such as fraud detection and medical illness detection. Some consider tasks like summarization and semantic content exploration, such as understanding the data, to be data mining tasks.

Here are some of our observations. While one can see differences between the different tasks, often there are similarities. When we teach a data mining course, the students frequently make statements like, what is the difference between affinity grouping and estimation? There is still no theory behind data mining where one can be precise with the notions and definitions. It is still rather adhoc. Furthermore, while say neural networks are good for clustering, one cannot make definite statements such as technique A is used for task X and technique B is good for task Y. It is largely trial and error and one gets better with experience. Therefore, unless you do a lot of data mining, it is difficult to determine what is best for a particular situation. Figure 7-1 illustrates the process of getting data mining outcomes.

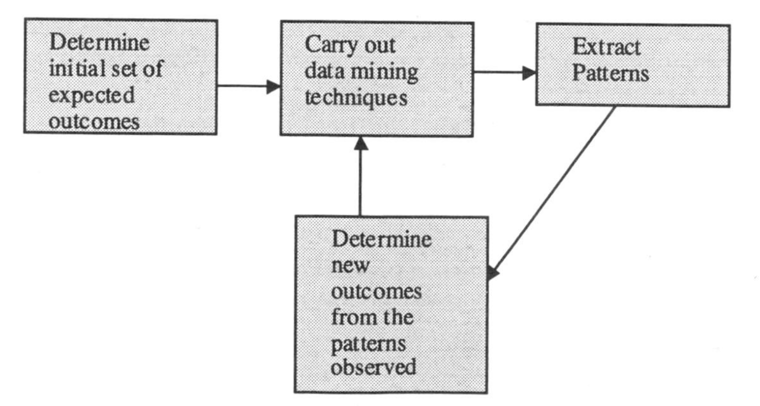

Figure 7-2. Data Mining Outcomes

7.3 APPROACHES TO DATA MINING

Berry and Linoff have [BERR97] clearly explained the approaches to data mining. They call it methodologies for data mining. These methodologies are not the outcomes nor are they the techniques. They are the steps one would take to do mining. Once the outcomes are determined then how do you go about doing the mining? Where do you start? This section addresses some of these issues.

Essentially there are two approaches: top-down and bottom-up. One can combine the two and have a hybrid approach. In the top-down approach you have to start with some idea or a pattern or a hypothesis. For example, a hypothesis could be "all those who live in Concord, Massachusetts earn a minimum of 50K." Then you start querying the database and test your ideas and hypothesis. If we find something that does not confirm our hypothesis, then we have to revise our hypothesis. A lot of statistical reasoning is used for this purpose. In general, hypothesis testing is about generating ideas, developing models, and then evaluating the model to determine if the hypothesis is valid or not. Developing the model is a major challenge. If the model is not a good one, then one cannot rely on the outcome. The models could simply be a collection of rules of the form "if a person lives in New York, then he owns a house worth more than 300K." To evaluate the model then, one needs to query the database. In the above ecample, one could query to select all those living in New York in homes costing less than 300K.

In the bottom-up approach to data mining, there is no hypothesis to test. This is much harder as the tool has to examine the data and then come up with patterns. The bottom-up approach could be directed or undirected. In directed data mining, also referred to as supervised learning in the machine learning literature, you have some idea what you are looking for. For example, who travels with John often to New York? What item is purchased often with milk? Like the top-down approach, models are developed and they are evaluated based on the data you analyze. With undirected data mining, also called unsupervised learning in the machine learning literature, you have no idea what you are looking for. You ask the tool to find something interesting. For example, in image data mining, the data mining tool can go about finding something that it thinks is unusual. As before, you develop a model and evaluate the model with the data. Once something interesting is found, then you can conduct directed data mining.

The hybrid approach is a combination of both top-down and bottom-up mining. For example, you can start with bottom-up mining, analyze the data and then discover a pattern. This pattern could be a hypothesis and you can do top-down mining to test the hypothesis. As a result, you can find new patterns which become a new hypothesis. That is, the tool can switch between top-down and bottom-up mining and again between directed and undirected mining. Data mining approaches are illustrated in Figure 7-3.

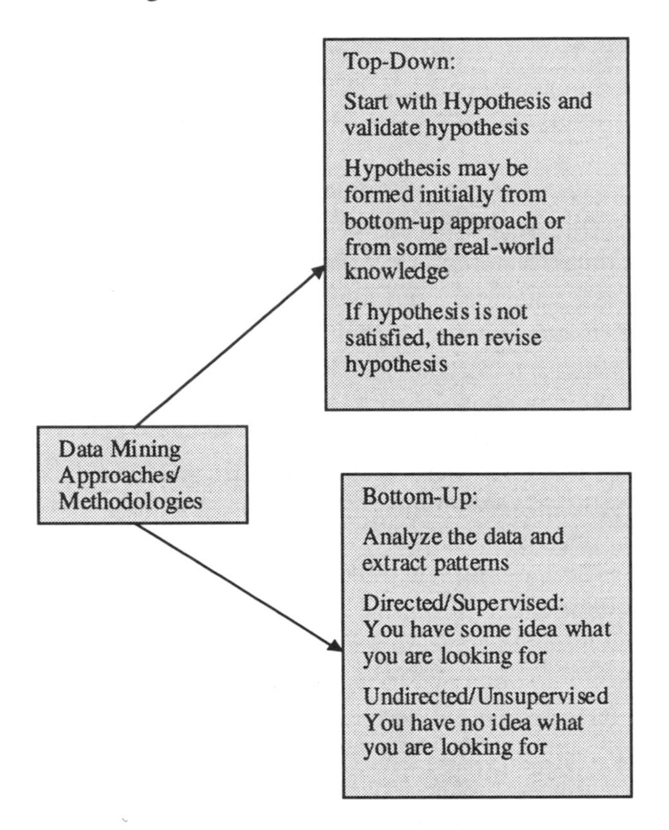

Figure 7-3. Some Data Mining Approaches

7.4 DATA MINING TECHNIQUES AND ALGORITHMS

Now we come to the important part of data mining, and that is the algorithms and techniques employed to do the mining. Data mining techniques are numerous. These include statistical analysis, machine learning, and other reasoning techniques. We discuss only a few of them. It should be noted that we have not distinguished between techniques and algorithms. One can argue that while techniques describe a broad class of procedures to carry out mining, algorithms go into more details. For example, while link analysis can be regarded to be a data mining technique, one could employ various algorithms, such as intelligent searching and graph traversal, to carry out link analysis. The outcome of link analysis would be to make associations between various entities.

One popular class of data mining techniques has come to be called market basket analysis. These are techniques that group items together. For example, which items go together, who travels with whom together, and what events occur together? The actual techniques employed to do market basket analysis are intelligent searching and pruning the search. Many of the intelligent search techniques that were developed for artificial intelligence are being employed for market basket analysis. If one were to search the entire search space, then it will be combinatorially explosive. Therefore, the challenge here is to determine how to search and eliminate unnecessary items from the search space. Various papers and books have given examples of market basket analysis from supermarket purchases. The idea is to make a list of all purchases for a certain period and then analyze these purchases to see which items are often purchased together. There will be some obvious answers such as bread and milk or bread and cheese. What the decision maker is looking for is some of the non-obvious answers like bread and soy sauce.

Another data mining technique is decision trees. This is a machine learning technique and is used extensively for classification. Records and objects are divided into groups based on some attribute value. For example, the population may be divided into two groups, one consisting of those who earn less than 100K and the other consisting of those who earn more than 100K. Each of the groups is then divided further based on some value such as age. That is, each class is further divided into subclasses depending on whether the age is more then 50 or less than 50. Then each of the subclasses can be further divided depending on marital status. Subsequently, a tree structure is formed with leaves at the end; the decision tree is then used for training. Then as new data appears to be analyzed, the training examples are used to classify the data.

Neural networks are another popular data mining technique that has been around for while. A neural network is essentially a collection of inputs signals, nodes, and output signals. They are first trained with training sets and examples. Once the learning is over, new patterns are given to the network. The network then uses its training experience and analyzes the new data. It may be used for clustering, identifying entities, deviation analysis and various other data mining tasks.

Inductive logic programming is a machine learning technique that is of special interest to us because of our own research. It originated from logic programming. Instead of deducing new data from existing data and rules, inductive logic programming is all about inducing rules from analyzing data. It has theory behind it and uses a variation of the resolution principle in theorem proving for discovering rules. Since inductive logic programming is of interest to us, we have devoted the next chapter to this topic.

Several other data mining techniques are in use today. They include link analysis techniques, which are a collection of techniques to find associations and relationships between records; automatic cluster detection techniques, which are a collection of techniques to find clusters; techniques to find association rules that are similar to link analysis; and nearest neighbor techniques, which are a collection of techniques that analyzes new data based on its neighbors. For example, in the nearest neighbor techniques, if a situation has to be analyzed, then examine the database to see if there are neighbors with similar properties. Then make conclusions about the new situation. The techniques employ distance functions to determine the closeness between the data entities. It is assumed that points in space that are close together have similar properties. So for new data, compute its point in space based on some predefined computation. Then find out how close it is to known data points and then determine its properties. Association rule-based techniques are popular among the database researchers. These techniques essentially examine the data in the database and come up with associations between the entities. In many ways the techniques are similar to those used for link analysis.

Other data mining techniques include those based on genetic algorithms, fuzzy logic, rough sets, concept learning, and simple rule-based reasoning. For a detailed discussion of these techniques, we refer to [LIN97], [BERR97], and [MITC97]. As mentioned, there is some overlap between the different techniques that have been proposed. These techniques have been taken from statistics, data management, and machine learning. As we make progress toward integrating the various data mining technologies, we can expect more and more sophisticated techniques to be developed. Figure 7-4 illustrates the way data mining techniques operate.

7.5 SUMMARY

This chapter has explained the concepts in data mining. We started with a discussion of the data mining outcomes, which are also referred to as tasks. These tasks determine what we can expect from data mining. Do we want clustering, classification, or affinity grouping? Then we discussed approaches or methodologies for data mining. For example, once you have determined what outcomes you want, how do you go about doing data mining? Do you take the top-down approach where you start with a hypothesis and then do testing, or do you take a bottom-approach where you do not start with a hypothesis, or do you take a hybrid approach which is a combination of the two? In the case of the bottom-up approach, do you carry out supervised learning where you direct the learning process, or do you do unsupervised learning where you have no idea what you want to learn in the beginning and start looking for something interesting? Finally, we discussed various data mining techniques such as decision trees, neural networks, and inductive logic programming.

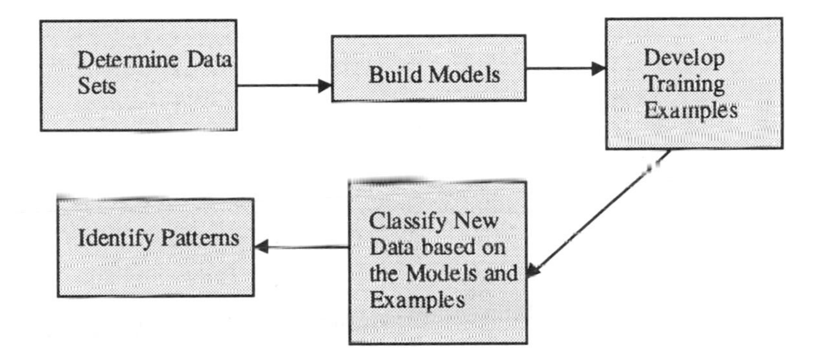

Figure 7-4. Operation of Data Mining Techniques

The concepts that we have discussed provide information on the essential points in data mining. We have tried to summarize a lot of information from tasks, approaches, and techniques into a single chapter. There are books devoted to various data mining techniques. We urge the reader to take advantage of the various material in the literature. Since inductive logic programming is a topic of interest to us, in the next chapter we discuss this technique. Then in Chapter 9 we provide an overview of the data mining tools and discuss the outcomes and techniques utilized by these tools.

CHAPTER 8

LOGIC PROGRAMMING AS A DATA MINING TECHNIQUE

8.1 OVERVIEW

In Chapter 7, we discussed various data mining techniques. One of the techniques we mentioned was inductive logic programming which has evolved from logic programming. Since this topic is of particular interest to us, as we have conducted research in logic programming (see, for example, [THUR90a]), we give special attention to inductive logic programming and data mining in this chapter.

Logic programming evolved as result of the pioneering work of people like Kowalski [KOWA74] when they developed the Prolog language. Much work was carried out in this area in the 1970s, and the database community and logic programming community started collaborating around the late 1970s. As a result, a lot of work was done on an area called deductive databases [GALL78]. Logic programming is essentially about using logic as a programming language. In many of the cases, the logic utilized is a subset of first order logic and is called Horn clause logic. Various programming languages based on higher order logic have been developed since then.

Deductive databases are databases that make deductions and inferences from rules and data. Deductive databases are essentially based on logic programming which has now come to be known as deductive logic programming [GALL78]. Significant developments in logic programming and deductive databases were made in the 1980s with the advent of the Japanese fifth generation project. This work has contributed much toward producing intelligent databases and programming systems.

In the late 1980s, interest began in an area called inductive logic programming where you learn rules from the data. That is, one extracts patterns and rules from the data. This is essentially knowledge discovery. While progress was made in inductive logic programming, knowledge discovery, more commonly known as data mining, started getting attention. The two fields started merging in the early 1990s. Today there is widespread acceptance of inductive logic programming (also now known as ILP) as a learning technique and, in particular, a data mining technique. There are now summer schools devoted to ILP for data mining (see, for example, [ILP97]).

This chapter describes the use of ILP as a data mining technique. To understand ILP, one needs to have an understanding of deductive logic programming and deductive databases. Therefore, to give some background we first provide an overview of deductive databases in Section 8.2. Then we discuss ILP in Section 8.3. A discussion of ILP and data mining is given in Section 8.4. Some ILP applications are discussed in Section 8.5. The chapter is summarized in Section 8.6.

8.2 DEDUCTIVE LOGIC PROGRAMMING

Deductive logic programming is the process of deducing data from existing data and rules. Deductive database systems are database systems based on deductive logic programming, and our discussion in this section will be on deductive databases since we are taking a data-oriented perspective in our discussion of data mining. We introduce this subject for the following reason. Within logic programming, while inductive logic programming is at one end of the spectrum, deductive logic programming system is at the other end. The former induces information while the latter deduces information.

Deductive database systems have also been referred to as logic database systems (or deductive logic programming systems). These are systems where the data model is based on the logic described in Chapter 2. Deduction rules are used to make new inferences. This way all of the data need not be stored in the database. That is, with the deduction rules, one could infer additional data. Deductive database systems was a very popular topic in the 1980s [ULLM88]. Work in this area resulted mainly from developments in databases, artificial intelligence, and logic programming research [GALL78. LLOY87]. Although it seems that these systems did not go beyond the prototype stage, the research on deductive and intelligent database systems has contributed to the developments in data mining to some extent. However, some applications of deductive database technology are given in [RAMA94].

Two of the architectures that have been examined by deductive database researchers are the loose coupling approach and the tight coupling approach (see, for example, [BROD86]and [DAS92]). With the loose coupling approach, the DBMS manages the database which is usually relational. The knowledge base consists of rules specified in a logic programming language such as Prolog [KOWA74]. The knowledge base management system (KBMS) manages the knowledge base. The inference engine component of the KBMS examines the rules and makes deductions. The DBMS is accessed to retrieve data. In the tight coupling approach, there is no separation between the DBMS and the KBMS. A module integrates the functions of a KBMS and a DBMS, and it also manages the knowledge base consisting of both the database and the rules. The architectures are illustrated in Figures 8-1 and 8-2.

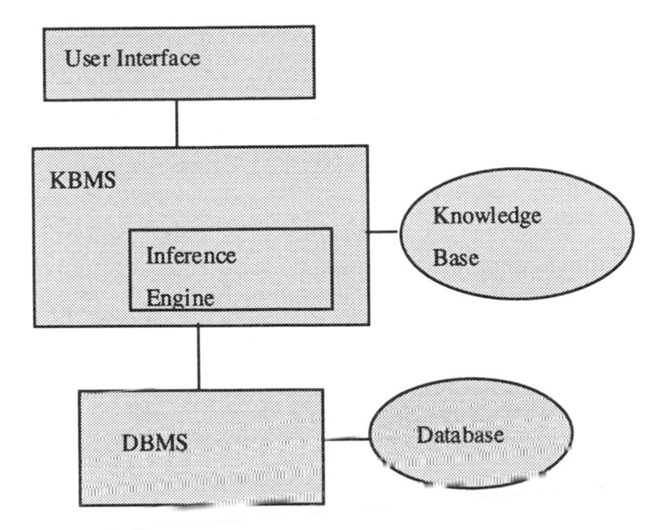

Figure 8-1. Loose Coupling Architecture

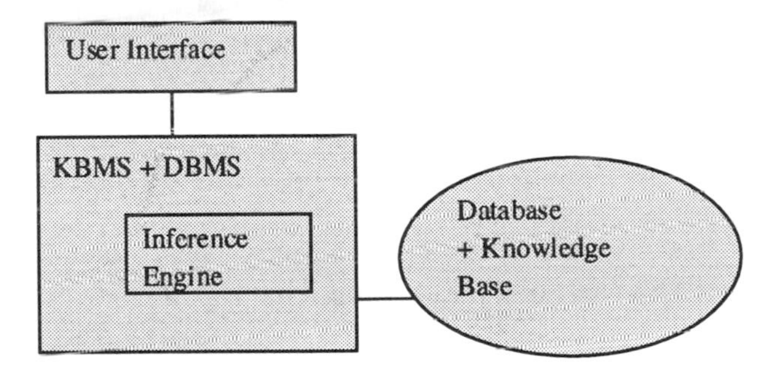

Figure 8-2. Tight Coupling Architecture

Note that we discussed logic as a data model in Chapter 2. We illustrate the concepts with a simple example which has been used in numerous cases. Suppose the database consists of the PARENT relation and there is a rule in the rule base that the parent of a parent is a grandparent. If the query is to retrieve all the (grandparent, grandchild) pairs, then the inference engine will examine this rule and query the database to get all pairs (X, Y) such that there is a Z where (X, Z) and (Z, Y) are in the PARENT relation. The resulting pairs (X, Y) is the response. Note that in this example, one does not have to explicitly store the relation GRANDPARENT in the database.

This section has provided only some of the essential points in logic programming and logic databases. For a detailed discussion, we refer to [LLOY87] and [GALL78]. Next we will move on to inductive logic programming and data mining.

8.3 INDUCTIVE LOGIC PROGRAMMING

While deductive logic programming infers data from data and rules, inductive logic programming (ILP) infers rules from the data. That is, ILP is all about inducing rules from the data. Since ILP is based on logic and inductive inference, it also has theory behind it. In the next few paragraphs we discuss ILP, and then in Section 8.4 we discuss its relationship to data mining.

Consider the parent relation discussed in the previous section. Suppose we have lots of data of the following form:

Parent(John, James) ←
Parent(James) ←
Grandparent(John, Mary) ←
Parent(Peter, Jane) ←
Parent(Jane, Bill) ←
Grandparent(Peter, Bill) ←

From all this data one could deduce the rule that the parent of a parent is a grandparent. That is, the rule is of the form:

Grandparent(X, Y) ← Parent(X, Z) and Parent(Z, Y)

Like in the case of deductive logic programming, in inductive logic programming one could also have an inductive inference engine that makes inductions. This inference engine could be a loosely coupled engine with the DBMS as illustrated in Figure 8-3, or it could be a tightly coupled engine as illustrated in Figure 8-4. ILP in its purest form carries out deterministic reasoning. That is, suppose the database has the following data.

Parent(Robert, Jill) ←
Grandparent(Robert, Kate) ←

Now suppose the database does not have the following data.

Parent(Jill, Kate) ←

Then the system will not deduce the Grandparent rule that we stated earlier. This is because there is not 100% support for this rule. Therefore, various extensions to ILP are being examined to support nondeterministic reasoning.

8.4 INDUCTIVE LOGIC PROGRAMMING FOR DATA MINING

We have seen that while deductive logic programming is all about deducing new data from existing data and the rules, inductive logic programming is all about learning rules from the data. Therefore, ILP is now regarded to be a branch of machine learning and subsequently a data mining technique.

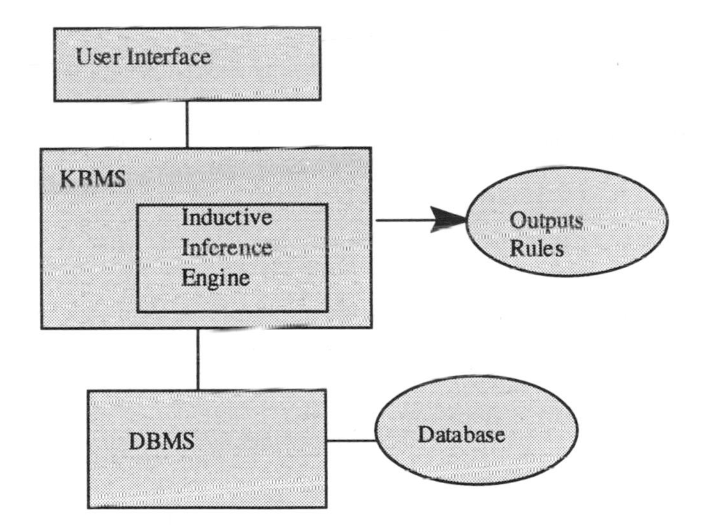

Figure 8-3. Loose Coupling Architecture

As mentioned earlier, there have been some criticisms that ILP will work only for small amounts of data and may fail for large quantities of data. Furthermore, ILP as a relational learning technique will work well for binary relationships and may not work well for more complex relationships. These are some of the criticisms that deductive logic programming has also received in the past.

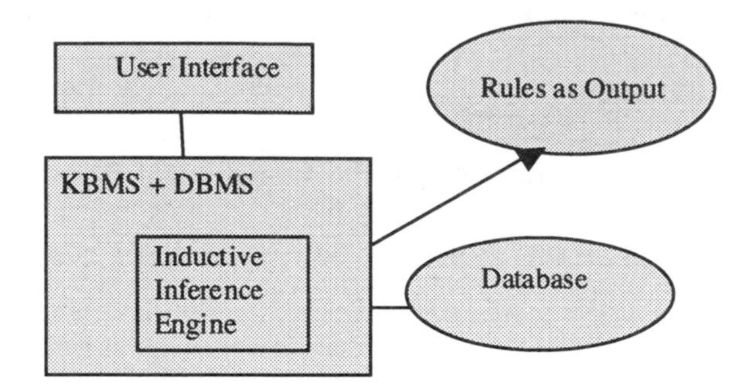

Figure 8-4. Tight Coupling Architecture

We believe that while some of these criticisms are valid, both deductive logic programming and inductive logic programming have contributed to the development of more intelligent systems. With respect to ILP, it is important to start with small databases and binary relationships and then move on to complex relationships. Furthermore, one has to have a good handle on the deterministic cases before tackling nondeterministic cases. For example, ILP works well with deterministic cases. If there are situations with data in the database where "Mary is the grandparent of Bob" and "Jane is the parent of Bob," but we do not have the data "Jim is the parent of Jane," then the rule "parent of a parent is a grandparent" cannot be learned. That is, even if there are many other cases that support it, if there is one exception, in general, ILP cannot learn such a rule. Therefore, ILP has been extended to support nondeterministic reasoning such as assigning probabilities and uncertainty values.

The main question we have is what technique does ILP use to learn? As discussed in various books and papers on ILP and machine learning (see, for example, [MITC97]), the resolution principle defined for theorem proving (see the discussion in [CHAN73]) has been extended to handle resolution in ILP. Through this resolution, ILP learns the rules. Another possibility is to keep a history of all the data, and as new data comes into the database, compare it to the previous data to see if there are similarities. However, this will not be efficient and is not based on sound principles. Therefore, much of the work on ILP is based on an aspect of resolution and theorem proving.

Once the rules have been learned, then what confidence do we assign to the rules? ILP in its purest form does not provide these facilities and also the capability to revise the rules. Additional reasoning techniques are needed for this purpose. However, since ILP is a learning technique, it has also become a technique for data mining. We need metrics and measurers to evaluate various data mining techniques and to see how well various ILP systems perform. Figure 8-5 illustrates the relationship between databases, logic programming, data mining, deductive database systems, machine learning, and relational learning.[14]

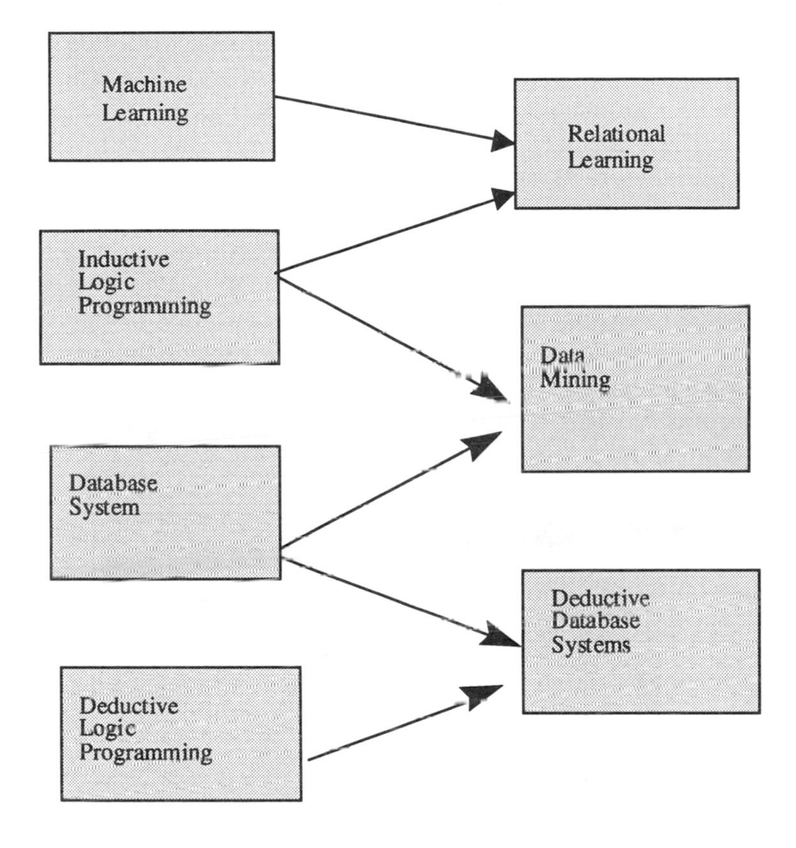

Figure 8-5. Data Mining, Database, and Logic Programming

8.5 APPLICATIONS OF INDUCTIVE LOGIC PROGRAMMING

While ILP is not an extremely active research topic in the United States, there are several initiatives on ILP with the European Community [DSV98] and there is also much interest in ILP in Japan. ILP is being applied to various applications. In particular, ILP is most suitable

[14] Relational learning is a special type of machine learning technique based on ILP and is about deducing relationships from the various data elements or entities.

for those applications where structure can be extracted from the instances. In this section we discuss a few.

One application where ILP works well is generating schemas from the data elements. This is a fairly deterministic application. Essentially, this is about generating the structure of the relational database from the data elements in the relations. ILP could also be used to generate the structure of the objects and the relationships between objects from the various objects specified. However, since schemas for objects are more general and nondeterministic, ILP may not be the best approach. For example, a document may have section components. While document A may have 3 sections, document B may have 4 sections. That is, the number of sections in a document is not deterministic.

ILP is also being applied to various projects in biochemistry and genetics [DSV98]. For example, in the case of biochemistry, ILP can be applied to extract the structure of the chemical compounds and in the case of genetics, ILP may be applied to extract the structure of the human g-nomes. While there has been much research on ILP, it is only recently that ILP is being applied to various application areas. Figure 8-6 illustrates the various application domains for ILP.

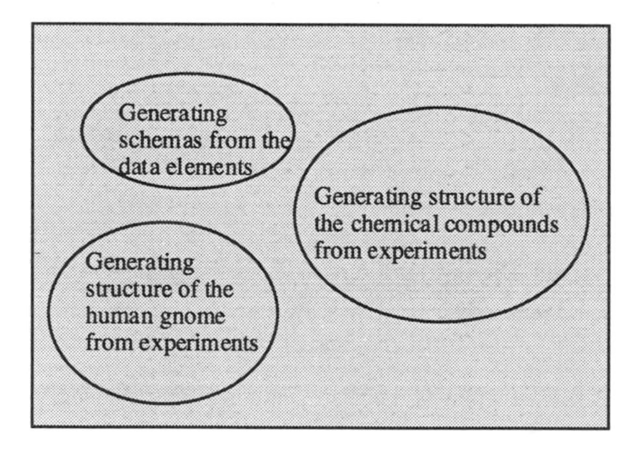

Figure 8-6. Some Application Areas for ILP

8.6 SUMMARY

This chapter has discussed an interesting data mining technique called inductive logic programming. It started with some background information on deductive logic programming and then discussed the essential points of inductive logic programming. The use of ILP as a data mining technique and its limitations were discussed.

Whether ILP will become a successful data mining technique remains to be seen. However, with solid theory behind it, ILP is very useful for deterministic data mining. In time, with appropriate extensions to ILP, we will see practical applications of ILP increase. Nevertheless, ILP is a very interesting technique that can provide much insight into machine learning and data mining.

We have found ILP to be useful in our study of data mining because it is based on logical reasoning. ILP has also helped us to have a better understanding of machine learning. This could be partly due to the fact that we have carried out fairly extensive work in logic programming systems, especially applied to secure database management, and therefore, learning ILP for data mining is a natural consequence.

For the beginner in data mining, without prior knowledge of logic programming, learning ILP might not be straightforward. Nevertheless, we encourage the reader to gain some knowledge of ILP as it would be helpful in understanding the logical reasoning behind data mining. We also believe that data mining is now very much an art. To make it into a science, we need more work in areas like ILP. This was also one of the main motivations for us to introduce ILP in this book.

CHAPTER 9

DATA MINING TOOLS

9.1 OVERVIEW

This chapter describes some example commercial data mining products and research prototypes, some of which have evolved into products. We group them into various categories as discussed in [CLIF96a]. As stated earlier, all of the information on these products has been obtained from published material as well as from vendor product literature. Since commercial technology is advancing rapidly, the status of these products as described here may not be current. Again, our purpose is to give an overview of what has been out there recently and not the technical details of these products.

We discuss only some of the key features of the commercial products and prototypes. Note that various data management texts and conferences including data management/mining magazines, books, and trade shows such as Database Programming and Design (Miller Freeman Publishers), Data Management Handbook Series (Auerbach Publications), and DCI's Database Client Server Computing Conferences have several articles and presentations discussing the commercial products. We urge the reader to take advantage of the information presented in these magazines, books, and conferences and keep up with the latest developments with the vendor products. Furthermore, in areas like world wide web mining, we can expect the developments to be changing very rapidly. The various web pages are also a useful source of information.

It should also be noted that we are not endorsing any of these products or prototypes. We have chosen a particular product or prototype only to explain a specific technology. We would have liked to have included discussions of many more products and prototypes. But such a discussion is beyond the scope of this book. In recent years various documents have provided a detailed survey of data mining products and prototypes. An example is the document on data mining products by the Two Crows Corporation. This corporation periodically puts out detailed surveys of the products and we encourage the reader to take advantage of such up-to-date information. There are also tutorials on comparing the various products (see, for example, [KDT98]).

We have divided the information on tools into two categories. One is the prototype tools that have emerged from research projects, and the other is commercial tools from vendors. In Section 9.2 we discuss the prototype tools, and in Section 9.3 we discuss the commercial products. The chapter is summarized in Section 9.4.

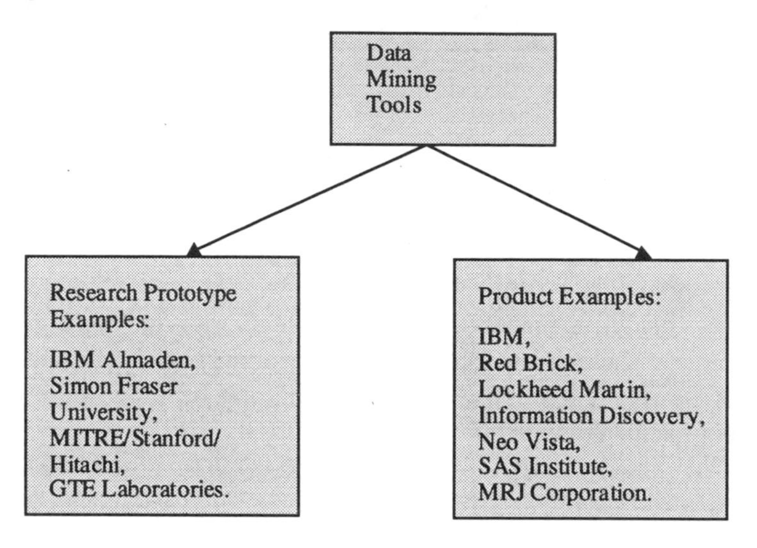

Figure 9-1. Classification of Data Mining Tools

9.2 PROTOTYPE TOOLS

9.2.1 Overview

Over the past decade numerous prototype data mining tools have emerged from various universities and research laboratories all over the world. Some of these tools are now commercially available. A discussion of all of these tools and systems is beyond the scope of this book. What we have done here is discuss the tools with respect to some special features they offer. Therefore, we have formed the following five groups for the tools. They are based on the following.

- New functional models
- New information services
- Scalability
- Understandability of the results
- Large-scale research projects

Our discussion has been influenced by the work of Clifton and Morey [CLIF96a] in surveying the various tools as well as the work by Grupe and Orang [GRUP98]. In the following subsections we will discuss the tools for each of the five areas that we have identified. Note that during the past year numerous other prototypes have emerged. A discussion of all of them is beyond the scope of this book. Some interesting systems include the systems developed at Northwestern University, and also at Imperial College, London. Research papers on these systems have appeared in [KDD98].

9.2.2 New Functional Models

The tools that fall under this category attempt to develop new models for data mining. In particular, frameworks for data mining are developed by the projects involved in developing new functional models. Example projects include one by Stanford University, MITRE Corporation, and Hitachi Corporation and another by Rutgers University.

The projects that attempt to develop new functional models essentially integrate data mining and database management. In particular the tight integration discussed in Chapter 2 are taken by these initiatives. The project by Stanford University, the MITRE Corporation, and Hitachi Corporation is called Queryflocks [TSUR98]. The idea here is to develop a query methodology and optimization techniques to handle flocks of queries to support data mining. In particular, these queries attempt to produce associations between entities in the database.

The project by Rutgers University also integrates data mining with database management and has formulated query languages for data mining queries. This project also attempts to find associations between the entities. A discussion of this work is given in [ACM96a].

9.2.3 New Information Services

The projects that work on new information services essentially mine different types of data such as multimedia data. That is, multimedia data mining is the focus of the projects that fall under this category. We discuss some of them here.

Data mining on text is being attempted by the Queryflocks project by Stanford University, the MITRE Corporation, and Hitachi Corporation. The technique for finding associations can be applied not only to relational databases but also to text databases once the tagged entities are extracted from the text. Another example of text mining is the work at the University of Arizona by Cheng and Ng. This project searches documents based on co-location of terms in documents. A third example is the work by Feldman at Bar-Ilan University in Israel that finds association rules between identified concepts in text.

There is some work also on image mining. Most notable is the SKICAT project by JPL (Jet propulsion Laboratory). This work detects unusual objects from images in space [ACM96a]. Another example of image mining work is that of Clifton et al., at the MITRE Corporation [CLIF98]. This work finds unusual patterns from hyperspectral images. There is also work on image mining at the University of British Columbia [NG97]. The technique used here is distance-based reasoning.

There is also some work on web mining that has been reported. Examples include those at the University of Michigan and at the University of Minnesota (see, for example, [ICTA97]).

9.2.4 Scalability

Scalability of the algorithms in data mining is still a largely unexplored area. The Massive Digital Data Systems Project has focused on scalability of various data management and data mining techniques to handle very large databases [MDDS94]. Scalability could be determined by using larger and larger data sets, by conducting theoretical studies as well as simulation studies.

Scalability of data mining algorithms needs a lot of work. There is some work at Magnify Inc. to determine the scalability of specific data mining techniques. These algorithms handle terabytes of data. Other products focusing on scalability of data mining techniques include those by Thinking Machines Corporation and SGI (Silicon Graphics). The work at IBM's Yorktown Heights research laboratory also addresses scalability issues.

Since data mining is still a relatively new area, there has not been much focus on scalability. However, as the techniques mature we can expect the focus to be on handling larger and larger data sets.

9.2.5 Understandability of the Results

In Chapter 6 we discussed the data mining process. After cleaning the data and mining the data, one needs to extract only the useful information. Therefore, understanding the data becomes very important. Some research projects focus on this aspect of data mining. We discuss a few.

GTE (General Telecommunications and Equipment) Laboratories has worked on data mining for a number of years [TKDE93]. They focus on the understanding of the data by producing domain specific reports. Medical cost mining is an application area for this work. Another effort is being carried out at Simon Fraser University. This is one of the prominent places for data mining. One aspect of this work is on integrating with visualization tools so that the data mining results can be understood better [HAN98]. A third effort is the work at the

University of Massachusetts at Lowell [GRIN97]. This work also focuses on integrating data mining with visualization techniques.

At present, much of the focus has been on applying data mining tools to extract patterns. Many say that this is what data mining is all about. Understandability of the results is the responsibility of some other area. However, if data mining is to be useful, we need to focus on understanding the data as well as mining the data.

9.2.6 Large-Scale Projects

Two very prominent projects in data mining include IBM's Quest project by Agarwal et al., and Simon Fraser University's *DBMINER* product by Han et al. Numerous papers have been published on this work (see, for example, [SIGM98]). We discuss only the essential points.

IBM's Quest project uses multiple data mining techniques and finds sequential associations as well as time-series associations. The work is influenced by database systems technology and builds data mining techniques to work with relational database systems. Some of this research has been transferred to IBM's products (to be discussed in Section 9.3). Simon Fraser University's DBMiner is now available as a product and focuses on mining relational data that has been warehoused and includes end-user support, visualization capabilities, and understanding of results. Mining association rules is the major focus of both the efforts.

The two large-scale projects that we have discussed here have influenced several other efforts that have emerged. Describing all of these efforts is beyond the scope of this book. However, recent conferences such as ACM SIGMOD conference, IEEE Data Engineering conference, the VLDB conference, as well as various data mining conferences we listed earlier contain many research papers that describe other emerging large-scale projects (see, for example, [SIGM98], [ICDE98], and [VLDB98]). Additional research papers in data mining can be found in [FAYY96].

9.3 COMMERCIAL TOOLS

9.3.1 Overview

As mentioned in Section 9.1, many data mining commercial tools are now emerging. This section provides an overview of some of the tools. We have selected these tools mainly because of our knowledge of them. As stated earlier, we are not endorsing any of these tools. The tools we will discuss are the following:

- Red Brick: *DATAMIND*
- Lockheed Martin: *RECON*
- IBM: *INTELLIGENT MINER*
- Information Discovery: *IDIS*
- Neo Vista: *DECISION SERIES*

Note that for each of the tools listed above, we have mentioned the corporation that develops the tool followed by the name of the tool.[15] Each of these products displays some interesting characteristics. The one by Red Brick integrates data warehousing with mining, the one by Lockheed focuses on mining general-purpose relational databases, the one by IBM is an example of company research being transferred to commercial products, the one by Information Discovery mines smaller relational databases particularly for the PC environment, and the one by Neo Vista develops a framework for mining.

Other data mining products include Whizsoft's *WHIZWHY* product, which is an end-user association rule-finding tool and uses rule-based reasoning. Hugin's product also called *HUGIN* uses Baysian reasoning and is good for prediction. *DATA LOGIC/R*, the product of Reduct Systems, uses rough sets as data mining technique, *NICEL* by Nicesoft uses fuzzy logic as a data mining technique. SGI's (Silicon Graphics) *MINESET* integrates data mining with visualization and focuses on high performance data mining. The product *DARWIN* by Thinking Machines Corporation also illustrates high performance data mining. SAS Institute's data mining product uses statistical reasoning quite heavily. SRA Corporation's product finds patterns for fraud detection. MRJ Corporation's data mining product does mining on large data sets. There are also many more products on the market, and as mentioned earlier in this chapter, describing all of them is beyond the scope of this book. A good tutorial comparing the various products was

---

[15] Note that many of these products are trademarks of their respective vendors. For example, Red Brick's warehousing product is called Red Brick™ Warehouse 5.0 and Lockheed's mining product is called RECON™. Because of the changes in trademarks due to acquisitions of corporations, we have not mentioned all of the trademarks explicitly in this book. For example, it was recently announced that Red Brick has been acquired by Informix Corporation. This means that the trade mark of Red Brick's products could change. If we know or have heard about a trademark for a product, then we have used capital letters in italics for the product when we mention it for the first time. We encourage the reader to keep up with the developments of the products from various web pages and vendor literature.

presented at the Knowledge Discovery in Databases Conference in 1998 [KDT].

9.3.2 Product 1

Red Brick's DataMind: Red Brick Corporation specializes in data warehousing. Recently they have developed a product called DataMind in collaboration with DataMind Corporation that tightly integrates the data warehousing and data mining products. We will first discuss Red Brick's data warehousing and then provide an overview of the mining product used for on-line analytical processing.

As quoted by Red Brick (private communication in 1996), Red Brick Warehouse is an open, relational, but specialized database designed from the ground up to address the data warehousing marketplace. The product's key advantages stem from the integration of very sophisticated join, indexing, and parallel processing technologies all focused on addressing the performance, functionality and scalability needs of data warehousing environments. Integration is a key aspect, as the Red Brick Warehouse architecture provides fast, consistent response times to complex queries against large datasets. Other features include subquery analysis in nested queries as well as returning initial rows to a query so that analysts can work with them while the remaining rows are being returned. In addition, parallel query processing as well as schema changes are also supported.

Red Brick Intelligent SQL has been a feature since the first version of Red Brick (version 1.5) in 1991. These business analysis extensions were incorporated to help answer the questions analysts ask about their corporate data that are difficult or impossible to ask in standard SQL. Providing this functionality is in keeping with Red Brick's philosophy that one needs a specialized database engine for data warehousing.

Data mining is part of the 5.0 release and is integrated as a simple extension of SQL. To utilize the data mining technology requires learning some new, simple SQL syntax. Red Brick has stated that integrating the data mining process with warehousing is a better approach than applying stand-alone data mining tools.

9.3.3 Product 2

Lockheed Martin's RECON: In this section we briefly discuss the product by Lockheed Martin Inc. called RECON. It should be noted that several research papers have also been published on RECON.

As mentioned by Lockheed (private communication 1995), RECON works on relational database systems such as those developed by Oracle Corporation, Sybase Inc., and Informix Corporation. It supports both data exploration through bottom-up data mining as well as pattern validation through top-down data mining. For example, an analyst can hypothesize patterns such as all those who live in Manhattan own cars worth more than 20K. RECON helps the analyst to validate the hypothesized patterns against the database. In the case of data exploration, RECON finds patterns previously unknown. Various deduction techniques are used for data exploration.

The application domains for RECON include stock portfolio creation and analysis, portfolio trading, loan risk analysis, credit analysis, marketing data analysis, and retail data analysis.

9.3.4 Product 3

IBM's Intelligent Miner: A popular data mining product on the market is IBM's Intelligent Miner. This product incorporates some of the research that has come out of IBM's Almaden Research Center. In particular, the Quest research project at Almaden has some of the origins of the Intelligent Miner product.

The techniques used by this product are many. The Intelligent Miner in a way is a multi-strategy data miner. In particular, it uses association rules, decision trees, neural networks and nearest neighbor methods for mining. It selects the methods as appropriate for the particular task. The outcomes it handles include missing values, anomaly detection, and categorization of continuous data, and it has applications in various domains.

One of the conditions for this product is that the data has to reside in IBM's database system product called *DB2*. However, this could change with time. The product is available for PC as well as mainframe environments.

9.3.5 Product 4

Information Discovery's IDIS: Information Discovery's IDIS is one of the earlier data mining tools running on Microsoft's *WINDOWS* as well as *NT* environments and is an end-user information discovery tool. It provides natural language reports and uses various induction and machine learning techniques. It operates on smaller relational databases.

Producing natural language reports of the data mining activities will help to understand the data mining results. Essentially, IDIS hypothesizes rules and then tests to see if the hypothesis is valid. It outputs unusual patterns as well as patterns that deviate from the norm. It has been used for a variety of applications in fields such as financial, medical, scientific research, and marketing.

9.3.6 Product 5

Neo Vista's Decision Series: Neo Vista's Decision Series product attempts to provide a framework for integrating multiple data mining products. It supports several data mining techniques such as association rules, neural networks, nearest neighbor algorithms, and genetic algorithms. This framework has tight integration with ODBC on the server side. ODBC provides the glue for integrating the various data mining products. The data miners can access data from multiple relational database systems which are integrated via ODBC.

Neo Vista's product essentially focuses on middleware for data mining that we have discussed in Chapter 5. This is one of the first products to provide such capability. We need this capability if multiple data mining strategies and products are to be integrated to develop more sophisticated data mining tools.

9.4 SUMMARY

This chapter has provided a brief overview of the various data mining tools including both research prototypes and commercial tools. These tools describe the various techniques utilized as well as the outcomes produced. In describing the research tools we chose five categories such as new functional models, new information services, understandability of the results, scalability, and large-scale projects. Then we described projects in each category. In describing commercial products, we selected five products and explained the essential points for each of them. We have selected these products only because of our knowledge about them. As mentioned earlier, we are not endorsing any of the products. Furthermore, due to the rapid developments in the field, the information about these products may soon be outdated. Therefore, we urge the reader to take advantage of the various commercial and research material available on these products.

The developments in data mining over the last few years have shown a lot of promise. Although some of the products have been around for a while, they are now being integrated with databases. As mentioned previously, we need the integration of multiple technologies to make data mining work. Furthermore, having good data is critical. Therefore in the future we will see more and more mining tools being integrated with various types of database systems as well as warehouses.

Conclusion to Part II

Part II described the essential points in data mining including the steps to data mining, the techniques, and tools used for mining. We also discussed a technique of special interest to us and that is inductive logic programming.

Note that the information we have provided in Part II relies on the underlying technologies we have discussed in Part I. That is, Part II builds on the data mining technologies. For example, many of the techniques we have discussed in Chapter 7 are based on machine learning. The steps involved in data mining clearly show the importance of having good data to mine. Database systems and data warehouses play a role here. We also addressed some of the difficulties in carrying out mining. We will revisit these challenges in Chapter 15 which concludes this book.

Now that we are familiar with what data mining is all about and the technologies that support data mining, we are now ready to explore the directions and trends. This is the subject of Part III.

Part III

Data Mining Trends

Introduction to Part III

While the previous two parts discussed technologies, techniques and tools, in this part we discuss trends and directions in data mining. In particular, some of the emerging activities in data mining are the focus of this part.

Part III consists of Chapters 10, 11, 12, 13, and 14. Chapter 10 describes data mining on distributed, heterogeneous and legacy databases. Much of the data is now distributed. Furthermore, heterogeneous data sources are being integrated. Finally, legacy databases are being migrated to new architectures. Data in these distributed, heterogeneous, and legacy databases have to be mined. Chapter 10 addresses some of these issues.

Chapter 11 describes mining multimedia data such as text, images and video. More and more data are now available in multiple media. Techniques for mining this data and extracting useful information are needed. Chapter 12 discusses web mining. Various database system vendors are providing access to the web. There is also structured and unstructured data on the web. This data has to be mined so that only relevant information is given to the decision maker. Furthermore, the usage patterns on the web have to be mined also so that a person browsing is given some advice.

Chapter 13 describes security and privacy issues. While all of the previous chapters in this book address the benefits of mining, this chapter discusses the security threats due to mining. With the mining tools, users now have a way of making deductions, and therefore security and privacy of various individuals could be compromised. Finally, Chapter 14 describes metadata and mining. Metadata is a useful resource for mining. On the one hand, the data miner could get the necessary information from the metadatabase in order to carry out its functions. On the other hand, the metadata itself could be mined to extract patterns.

CHAPTER 10

MINING DISTRIBUTED, HETEROGENEOUS, AND LEGACY DATABASES

10.1 OVERVIEW

Much of the discussions in the previous chapters assumed that a single database or warehouse has to be mined. Even in the case of multiple data sources we assumed that these data sources had to be integrated into a warehouse so that the warehouse could be mined. In many situations one would want to leave the data in the heterogeneous data sources and then mine these data sources. Furthermore, for many applications the data could be distributed and managed by a distributed database system. For such applications, the data mining tools have to operate on the distributed databases. Finally, there is still a lot of data residing in legacy databases. The major challenge is in mining and extracting useful information from these legacy databases. These legacy databases could also be migrated to new systems and architectures. So is it worth developing mining tools to operate on the legacy databases?

This chapter discusses issues and challenges on mining data in distributed, heterogeneous, and legacy databases. We first provide an overview of distributed databases, heterogeneous database integration, and migrating legacy databases. This overview is provided in Section 10.2. Then in Section 10.3 we discuss mining multiple data sources. First, the challenges on mining distributed databases are discussed, then we extend the discussion to mining heterogeneous data sources. Finally, mining legacy databases is discussed. The chapter is summarized in Section 10.4.

10.2 DISTRIBUTED, HETEROGENEOUS, AND LEGACY DATABASES

10.2.1 Distributed Databases

Although many definitions of a distributed database system have been given, there is no standard definition. Our discussion of distributed database system concepts and issues has been influenced by the discussion in [CERI84]. A distributed database system includes a distributed database management system (DDBMS), a distributed database, and a network for interconnection. The DDBMS manages the distributed database. A distributed database is data that is distributed across multiple databases. Our choice architecture for a distributed database system is a multi-database architecture which is tightly coupled. This architecture is illustrated in Figure 10-1. We have chosen such an architecture, as we can explain the concepts for both homogeneous and heterogeneous systems based on this approach. In this architecture, the nodes are connected via a communication subsystem and local applications are handled by the local DBMS. In addition, each node is also involved in at least one global application, so there is no centralized control in this architecture. The DBMSs are connected through a component called the Distributed Processor (DP). In a homogeneous environment, the local DBMSs are homogeneous while in a heterogeneous environment, the local DBMSs may be heterogeneous.

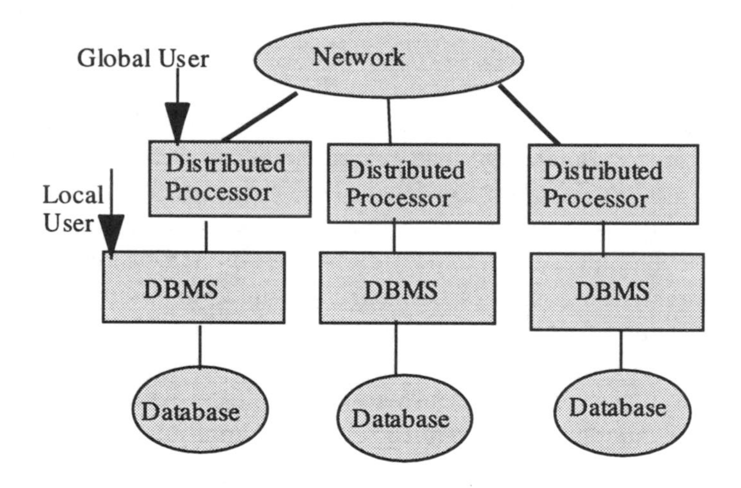

Figure 10-1. An Architecture for a DDBMS

Distributed database system functions include distributed query processing, distributed transaction management, distributed metadata management, and enforcing security and integrity across the multiple nodes [BELL92]. The DP is a critical component of the DDBMS. It is this module that connects the different local DBMSs. That is, each local DBMS is augmented by a DP. The modules of the DP are illustrated in Figure 10-2. The components are the Distributed Metadata Manager (DMM), the Distributed Query Processor (DQP), the Distributed Transaction Manager (DTM), the Distributed Security Manager (DSP), and the Distributed Integrity Manager (DIM). DMM manages the global metadata. The global metadata includes information on the schemas which describe the relations in the distributed database, the way the relations are fragmented, the locations of the fragments, and the constraints enforced. DQP is responsible for distributed query processing; DTM is responsible for distributed transaction management; DSM is responsible for enforcing global security constraints; and DIM is responsible for maintaining integrity at the global level. Note that the modules of DP communicate with their peers at the remote nodes. For example, the DQP at node 1 communicates with the DQP at node 2 for handling distributed queries.

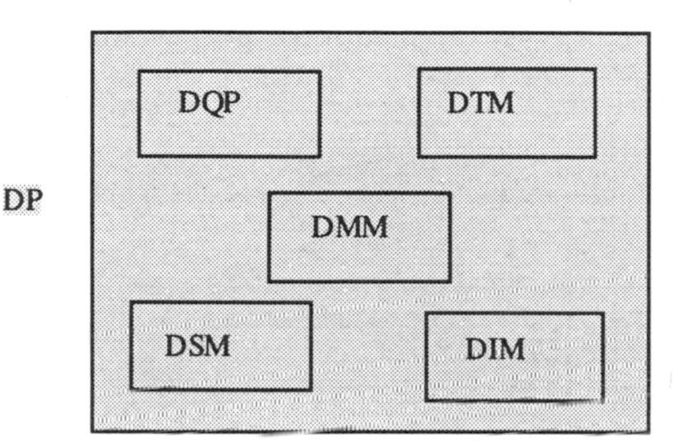

Figure 10-2. Modules of DP

10.2.2 Interoperability of Heterogeneous Database Systems

Figure 10-3 illustrates an example of interoperability between heterogeneous database systems. The goal is to provide transparent access, both for users and application programs, for querying and executing transactions (see, for example, [IEEE91], [ACM90], and [WIED92]). Note that in a heterogeneous environment, the local DBMSs may be heterogeneous. Furthermore, the modules of the DP have both local DBMS specific processing as well as local DBMS independent processing. We call such a DP to be a heterogeneous distributed processor (HDP). Some of these issues are discussed in more detail in [THUR97].

There are several technical issues that need to be resolved for the successful interoperation between these diverse database systems. Note that heterogeneity could exist with respect to different data models, schemas, query processing techniques, query languages, transaction management techniques, semantics, integrity, and security. There are two approaches to interoperability. One is the federated database management approach where a collection of cooperating, autonomous, and possibly heterogeneous component database systems, each belonging to one or more federations, communicates with each other. The other is the client-server approach where the goal is for multiple clients to communicate with multiple servers in a transparent manner. Our previous book on *Data Management Systems Evolution and*

Interoperation addresses both aspects to interoperability [THUR97]. Various aspects of heterogeneity are also addressed in that book. We are often asked the question as to when one should interconnect heterogeneous database systems through an HDP and when one should integrate them through a data warehouse? For on-line transaction processing applications the interoperability approach is the answer whereas for decision support applications the warehousing approach is the answer. For some other applications one would need both as illustrated in Figure 10-4.

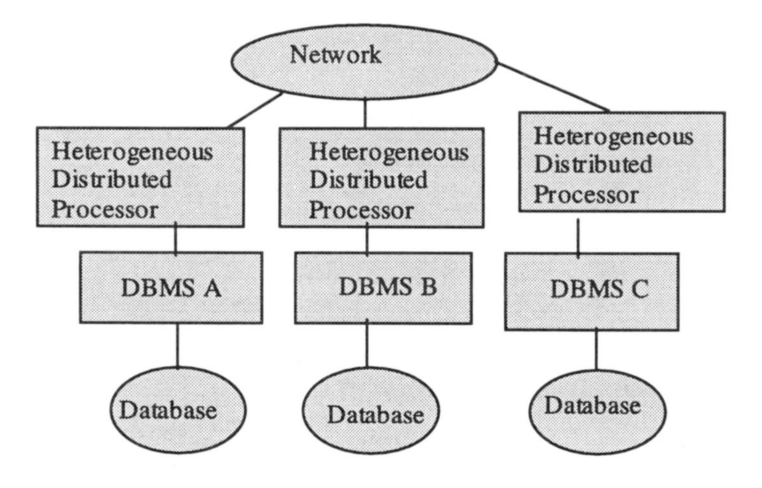

Figure 10-3. Interoperability of Heterogeneous Database Systems

10.2.3 Migrating Legacy Databases

Many database systems developed some twenty to thirty years ago are becoming obsolete. These systems use older hardware and software. Between now and the next few decades, many of today's information systems and applications will also become obsolete. Due to resource and, in certain cases, budgetary constraints, new developments of next generation systems may not be possible in many areas (see, for example, [BENS95]). Therefore, current systems need to become easier, faster, and less costly to upgrade and less difficult to support. Legacy database system and application migration is a complex problem, and many of the efforts underway are still not mature. While a good book has been published recently on this subject [BROD95], there is no uniform approach for migration. Since migrating legacy databases and applications is becoming a necessity for most organizations, both government and commercial, one could expect a considerable amount of resources to be expended in this area in the near future. The research issues are also not well understood.

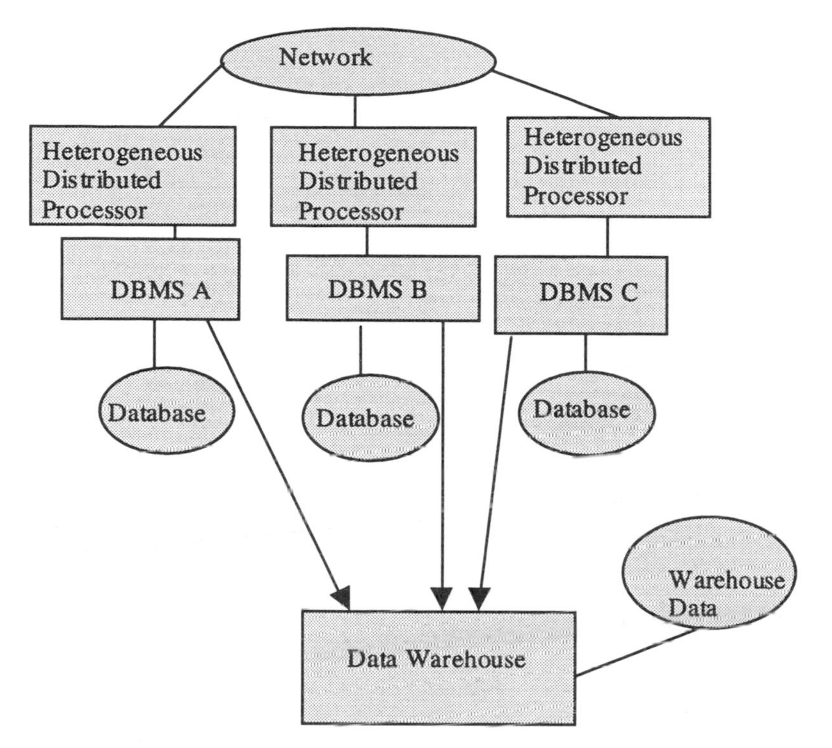

Figure 10-4. Interoperability and Warehousing

Migrating legacy applications and databases also has an impact on heterogeneous database integration. Typically a heterogeneous database environment may include legacy databases as well as some of the next generation databases. In many cases, an organization may want to migrate the legacy database system to an architecture like the client-server architecture and still want the migrated system to be part of the heterogeneous environment. This means that the functions of the heterogeneous database system may be impacted due to this migration process.

Two candidate approaches have been proposed for migrating legacy systems. One is to do all of the migration at once. The other is incremental migration. That is, as the legacy system gets migrated, the new parts have to interoperate with the old parts. Various issues and challenges to migration are discussed in [THUR97]. In the next section we will address how legacy databases could be mined. Figure 10-5 illustrates an incremental approach to migrating legacy databases through the use of object request brokers.

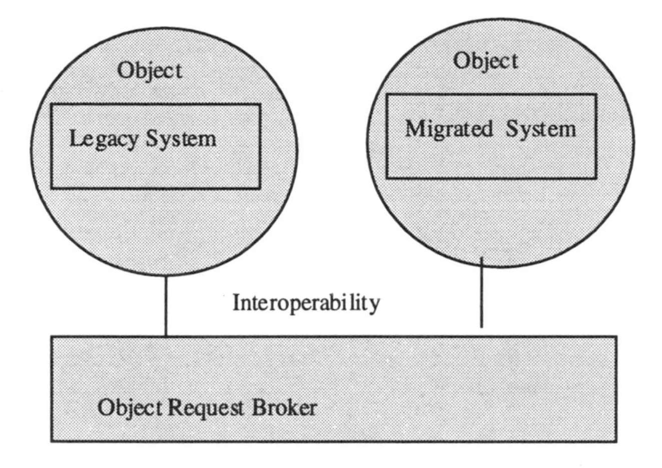

Figure 10-5. Migrating Legacy Databases

10.3 MINING DISTRIBUTED, HETEROGENEOUS, AND LEGACY DATABASES

In [THUR97], we placed much emphasis on heterogeneous database integration and interoperability. Many applications require the integration of multiple data sources and databases. These data sources may need to be mined to uncover patterns. Furthermore, interesting patterns may be found across the multiple databases. Mining heterogeneous and distributed data sources is a subject that has received little attention.

In the case of distributed databases, one approach is to have the data mining tool to be part of the distributed processor where each DP has a mining component also as illustrated in Figure 10-6. This way, each data mining component could mine the data in the local database and the DP could combine all the results. This will be quite challenging as the relationships between the various fragments of the relations or objects have to be maintained in order to mine effectively. Also, the data mining tool could be embedded into the query optimizer of the DQP which is the distributed query processing component of the DP. Essentially, with this approach, the DP has one additional module that is a distributed data miner. We call this module a DDM, as shown in Figure 10-7.

We illustrate distributed data mining with an example shown in Figure 10-8. Each DDM mines data from a specific database. These databases contain information on projects, employees and travel. The DDMs can mine and get the following information: John and James travel together to London on project XXX at least 10 times a year. Mary joins them at least four times a year.

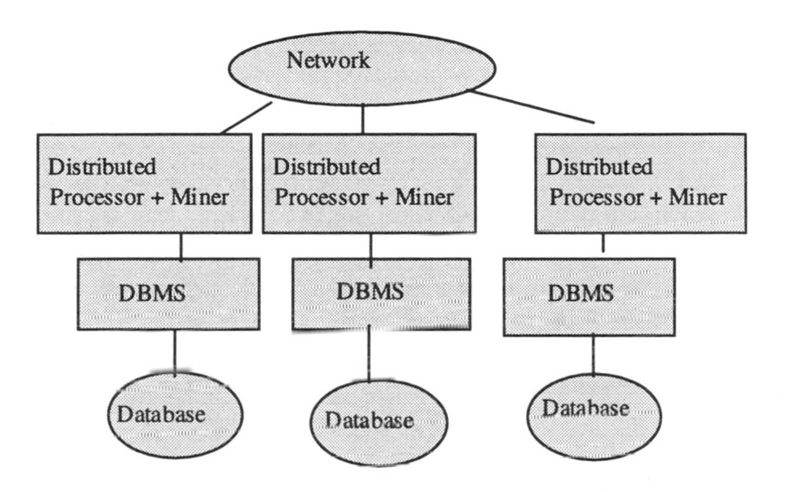

Figure 10-6. Distributed Processing and Mining

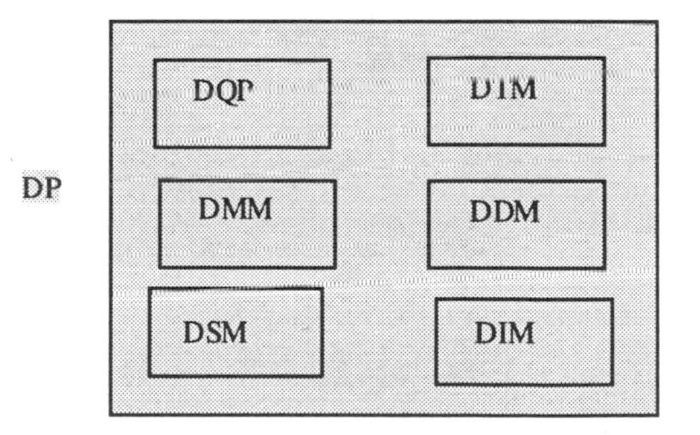

Figure 10-7. Modules of DP for Data Mining

An alternative approach is to implement the data mining tool on top of the distributed system. As far as the mining tool is concerned, the database is one monolithic entity. The data in this database has to be mined and useful patterns have to be extracted as illustrated in Figure 10-9.

In the case of heterogeneous data sources, we can either integrate the data and then apply data mining tools as shown in Figure 10-10 or apply data mining tools to the various data sources and then integrate the results as show in Figure 10-11. Note that if we integrate the databases first, then integration methods for interoperating heterogeneous databases are different from those for providing an integrated view in a distributed database. Some of these issues are discussed in [THUR97]. Furthermore, for each data mining query, one may need to first send that same query to the various data sources, get the results, and integrate the results as shown in Figure 10-10. If the data is not integrated, then a data miner may need to be integrated with the HDP as illustrated in Figure 10-11. If each data source is to have its own data miner, then each data miner is acting independently. We are not sending the same query to the different data sources as each data miner will determine how to operate on its data. The challenge here is to integrate the results of the various mining tools applied to the individual data sources so that patterns may be found across data sources.

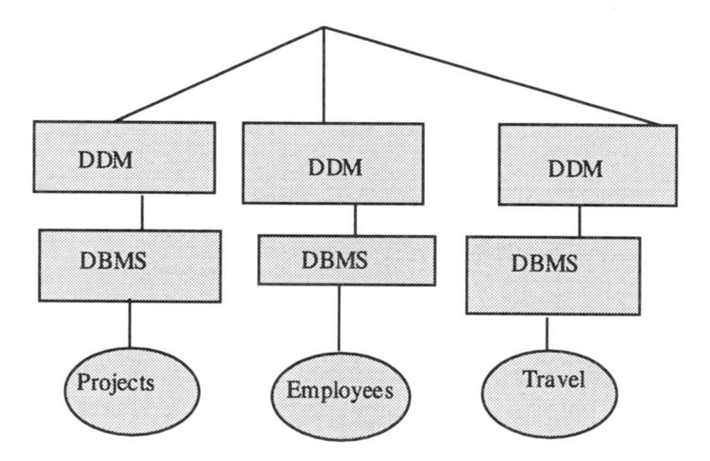

Figure 10-8. Example Distributed Data Mining

If one is to integrate the data sources and then apply the data mining tools, then the question is do we develop a data warehouse and mine the warehouse, or do we mine with interoperating database systems? Note that in the case of a warehouse approach, not all of the data in the heterogeneous data sources are brought into the warehouse. Only decision support data is brought into the warehouse. If interoperability is used together with warehousing, then the data miner could augment both the HDP and the warehouse as illustrated in Figure 10-12.

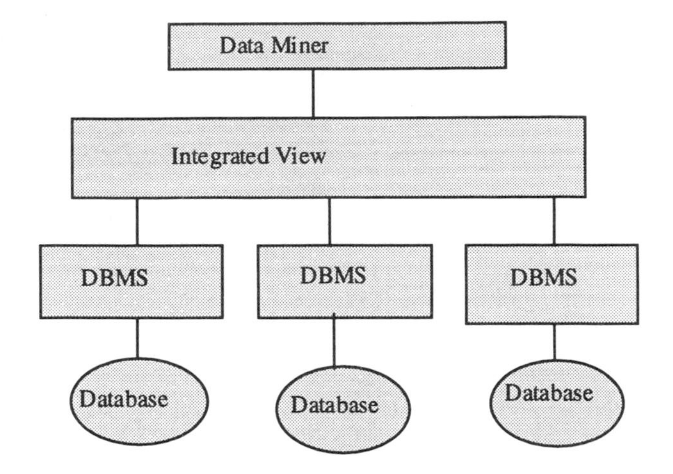

Figure 10-9. Data Mining Hosted on a Distributed Database

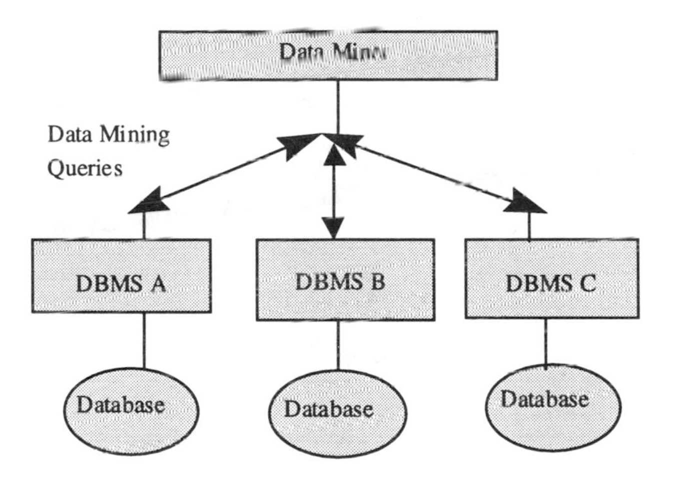

Figure 10-10. Data Mining on Heterogeneous Data Sources

One could also use more sophisticated tools such as agents to mine heterogeneous data sources as illustrated in Figure 10-13 where an integration agent integrates the results of all the mining agents. The integration agent may give feedback to the mining agents so that the mining agents may pose further queries to the data sources and obtain interesting information. There is two-way communication between the integration agent and the mining agents. Another alternative is not to have an integration agent, but to have the various mining agents collaborate with each other and discover interesting patterns across the

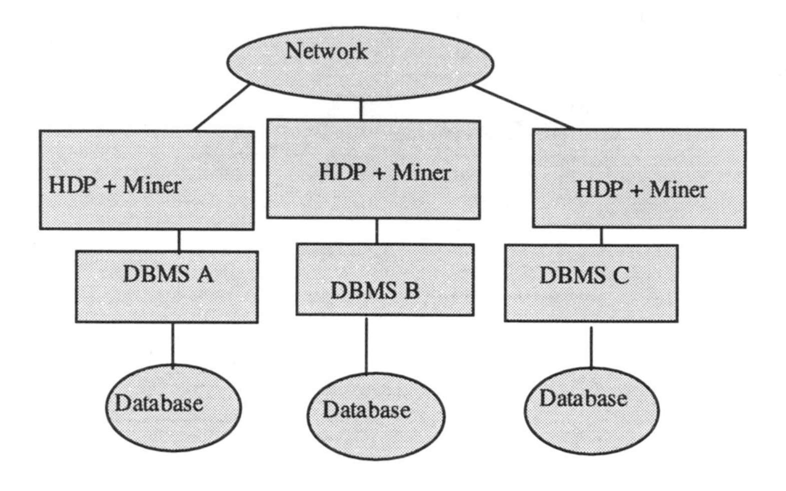

Figure 10-11. Mining and Then Integration

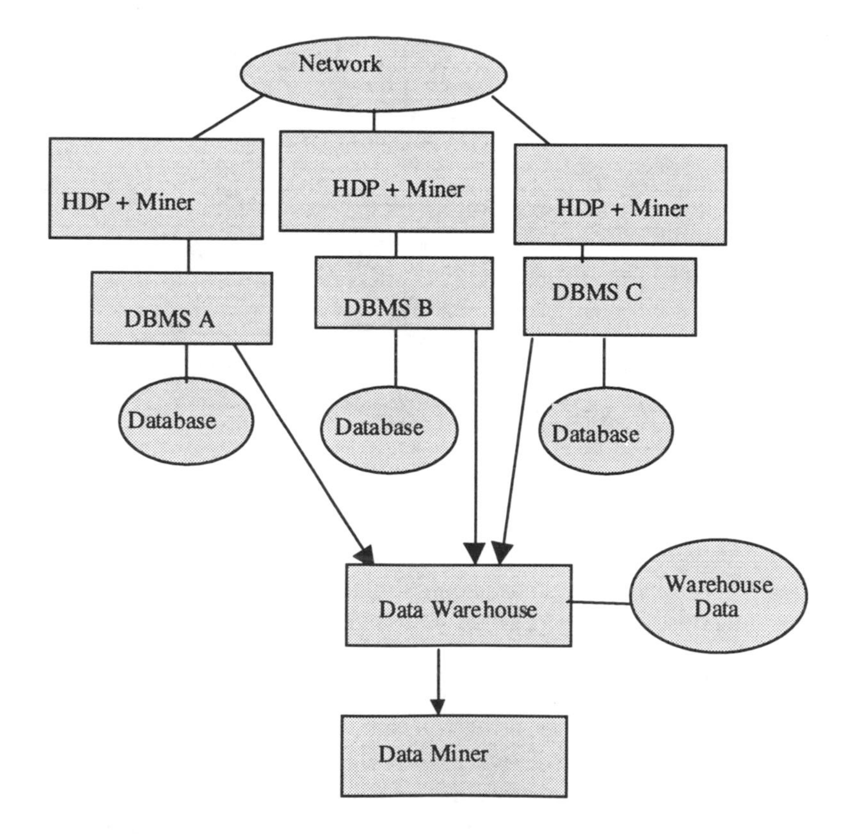

Figure 10-12. Mining, Interoperability, and Warehousing

various data sources. This is illustrated in Figure 10-14. This latter approach is also called collaborative data mining. In this approach collaborative computing, data mining, and heterogeneous database integration technologies have to work together.

The specific approach to mining heterogeneous data sources, whether to use an integration agent or have the mining agents collaborate, is yet to be determined. One may need both approaches or there may be yet another approach. Note also that heterogeneity may exist with respect to data models, data types, and languages. This could pose additional challenges to the data mining process. There is much research to be done in this area.

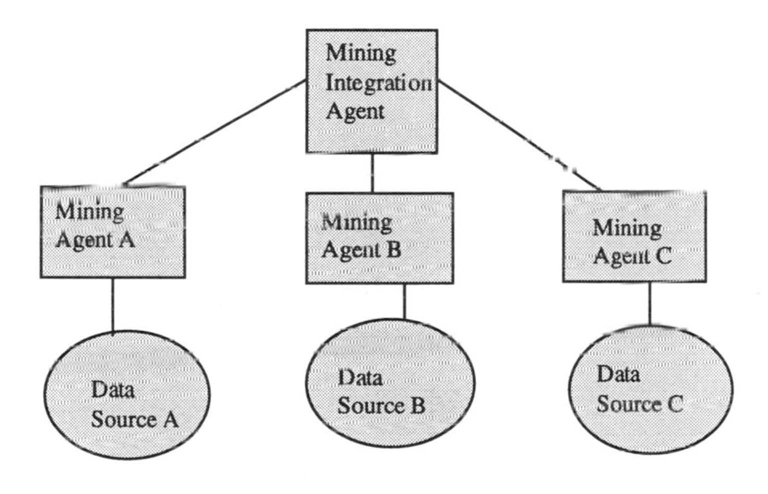

Figure 10-13. Integrating Data Mining Agents

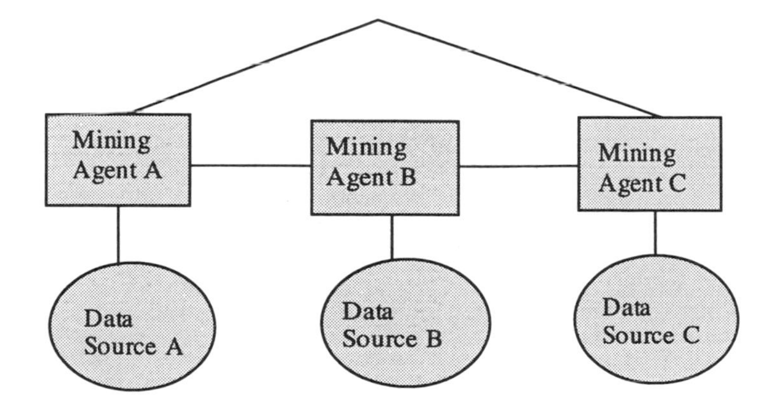

Figure 10-14. Collaboration Among Mining Agents

Another scenario for collaborative data mining is illustrated in Figure 10-15. Here two teams at different sites use collaboration and mining tools and mine the shared database. One could also use mediators to mine heterogeneous data sources. Figure 10-16 illustrates an example where we assume that there are general purpose data miners and mediators are placed between the data miners and the data sources. We also use a mediator to integrate the results from the different data miners.

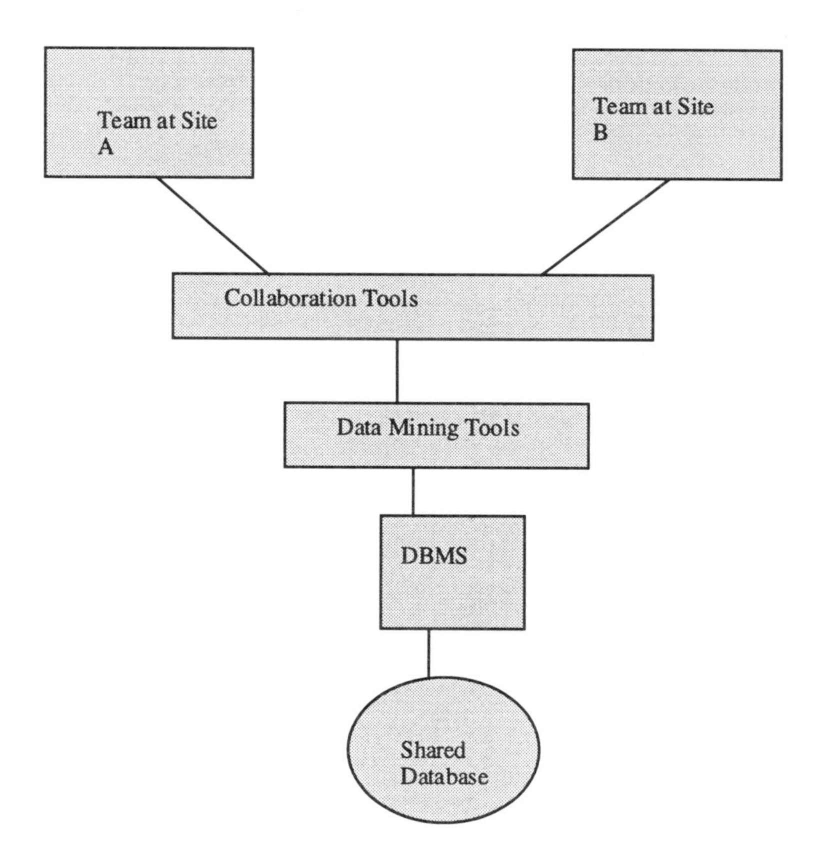

Figure 10-15. Teams Conducting Mining on Shared Database

Next, let us focus on legacy databases. One of the challenges here is how to mine the legacy databases? Can one rely on the data in these databases? Is it worth organizing and formatting this data especially if it has to be migrated to newer systems? Is it worth developing tools to mine the legacy databases? How easy is it to integrate the legacy databases to form a data warehouse? There are some options. One is to migrate the legacy databases to new systems and mine the data in the new systems (see Figure 10-17). The second approach is to integrate legacy databases and form a data warehouse based on new architectures and technologies and then mine the data in the warehouse (see Figure 10-18). In general it is not a good idea to directly mine legacy data, as this data could soon be migrated, or it could be incomplete, uncertain and therefore expensive to mine. Note that mining could also be used to reverse engineer and extract schemas from the legacy databases (see Figure 10-19), and may be applied to handle the Year 2000 problem. For a discussion of the Year 2000 problem we refer to [THUR97].

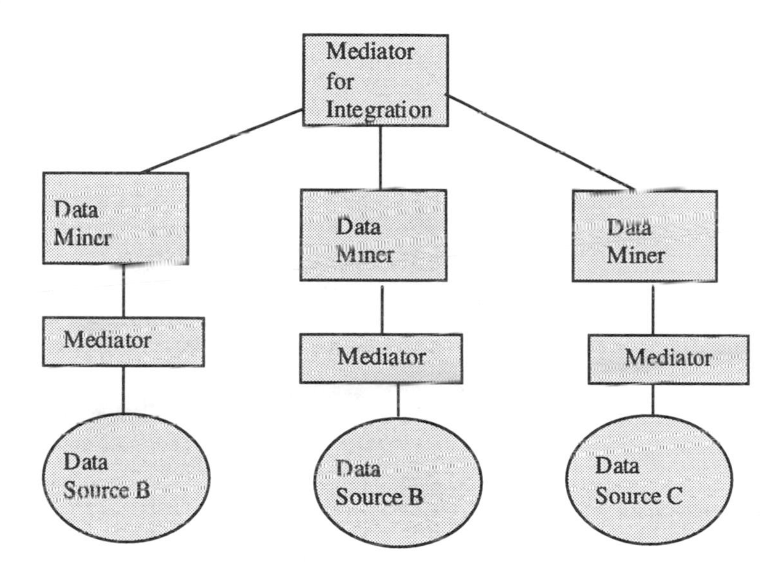

Figure 10-16. Mediator for Integration

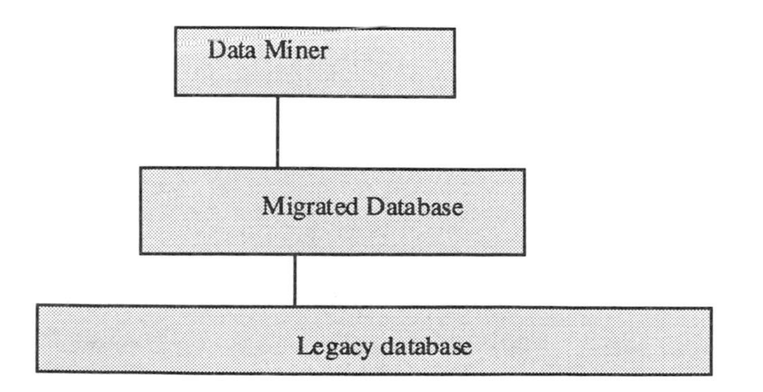

Figure 10-17. Migration and then Mining

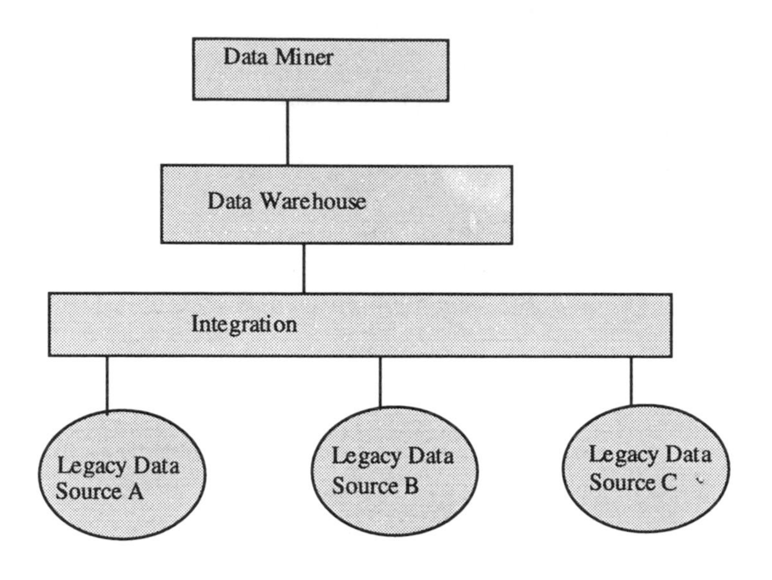

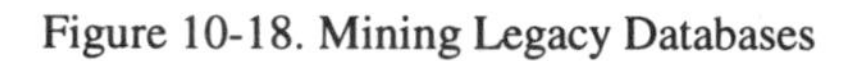
Figure 10-18. Mining Legacy Databases

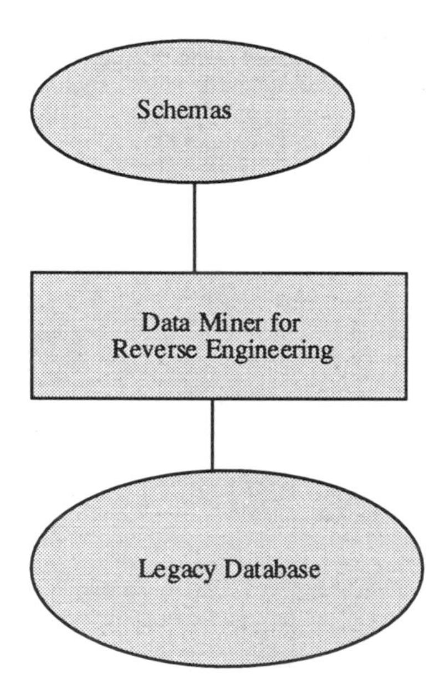

Figure 10-19. Extract Schemas from Legacy Databases

10.4 SUMMARY

In this chapter we have provided a brief overview of mining distributed, heterogeneous, and legacy databases. This is a relatively new research area and there is little progress to report. However, as data mining tools get more sophisticated, we expect them to be used on distributed, heterogeneous, and legacy databases. We have pointed out some of the issues and the work that needs to be done.

Recently there have been various workshops focusing on mining distributed databases [DIST98]. However, learning rules from the distributed data sources has received little attention. Furthermore, handling the various types of heterogeneity that we have discussed here as well as in our previous work (see, for example, [THUR97]) with respect to mining has received little attention. Mining multimedia databases, which handles a special type of heterogeneity, has received some attention. Some of the issues will be addressed in Chapter 11. We expect considerable progress to be made in mining distributed and heterogeneous databases in the future. As we have discussed, the approaches for heterogeneous database mining include integrating the data and then mining, which is also an approach for distributed databases; or mining the individual data sources with separate data miners and then integrating the results; or having agents collaborate with one another and conduct mining.

As far as legacy databases are concerned, the major issue is whether it is worth developing mining tools for such databases if they are to be migrated? Furthermore, these databases may be incomplete and of poor quality. Data warehousing could play a role in this. Data mining could also be applied to extract schemas from the legacy databases.

CHAPTER 11

MULTIMEDIA DATA MINING

11.1 OVERVIEW

A multimedia database system includes a multimedia database management system and a multimedia database. A multimedia database management system (MM-DBMS) manages the multimedia database. A multimedia database is a database, which contains multimedia data. Multimedia data may include structured data as well as semi-structured and unstructured data such as audio, video, text, and images. An MM-DBMS provides support for storing, manipulating, and retrieving multimedia data from a multimedia database. In a sense, a multimedia database system is a type of heterogeneous database system, as it manages heterogeneous data types. Heterogeneity is due to the media of the data such as text, video, and audio.

An MM-DBMS must provide support for typical database management system functions. These include query processing, update processing, transaction management, storage management, metadata management, security, and integrity. In addition, in many cases, the various types of data such as voice and video have to be synchronized for display, and therefore, real-time processing is also a major issue in an MM-DBMS.

Recently there has been much interest on mining multimedia databases such as text, images, and video. As mentioned, many of the data mining tools work on relational databases. However, considerable amount of data is now in multimedia format. There is lots of text and image data on the web. News services provide lots of video and audio data. This data has to be mined so that useful information can be extracted. One solution is to extract structured data from the multimedia databases and then mine the structured data using traditional data mining tools. Another solution is to develop mining tools to operate on the multimedia data directly. Multimedia data mining is the subject of this chapter. Note that to mine multimedia data, we must mine combinations of two or more data types, such as text and video, or text, video, and audio. However, in this book we deal mainly with one data type at a time. This is because we need techniques to mine the data belonging to the individual data types first before mining multimedia data. In the future we can expect tools for multimedia data mining to be developed.

In Section 11.2, we first provide some useful information on multimedia databases so the reader can understand the mining concepts better. In particular, architectures, modeling, and multimedia database functions are discussed. Then in Section 11.3, we discuss multimedia mining including text mining, image mining, video mining, and audio mining. We also briefly address the issues on mining combinations of data types. The chapter concludes with Section 11.5.

11.2 MULTIMEDIA DATABASES

11.2.1 Architectures for an MM-DBMS

Various architectures are being examined to design and develop an MM-DBMS. In one approach, the DBMS is used just to manage the metadata, and a multimedia file manager is used to manage the multimedia data. Then there is a module for integrating the DBMS and the multimedia file manager. This architecture is based on the loose-coupling approach and is illustrated in Figure 11-1. In this case, the MM-DBMS consists of the three modules: the DBMS managing the metadata, the multimedia file manager, and the module for integrating the two.

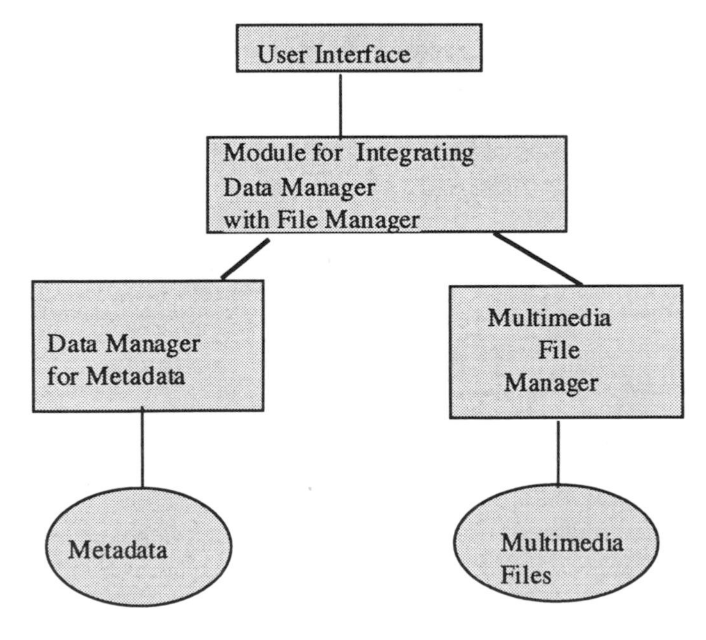

Figure 11-1. Loose Coupling Architecture

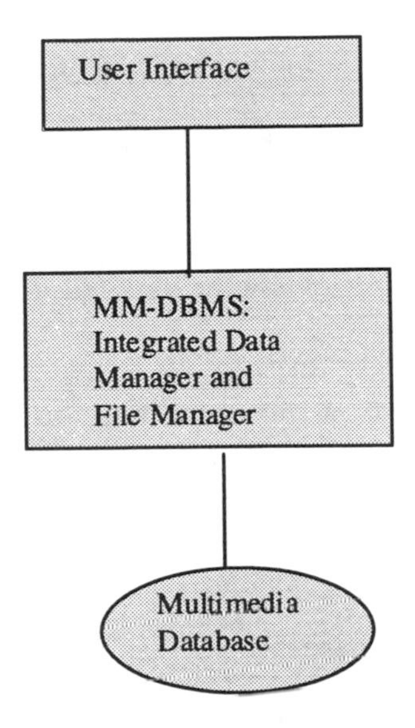

Figure 11-2. Tight Coupling Approach

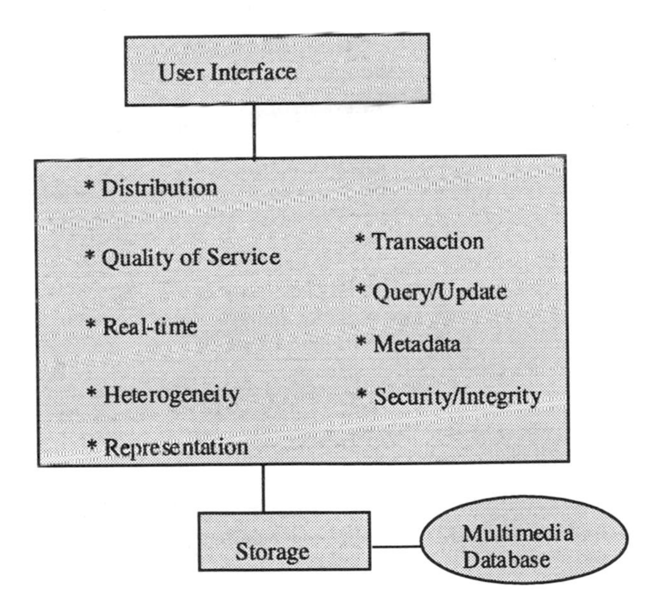

Figure 11-3. Functional Architecture

The second architecture, illustrated in Figure 11-2, is the tight coupling approach. In this architecture, the DBMS manages both the multimedia database as well as the metadata. That is, the DBMS is an

MM-DBMS. The tight coupling architecture has an advantage because all of the DBMS functions could be applied on the multimedia database. This includes query processing, transaction management, metadata management, storage management, and security and integrity management. Note that with the loose coupling approach, unless the file manager performs the DBMS functions, the DBMS only manages the metadata for the multimedia data. Functional architecture is illustrated in Figure 11-3.

There are also other aspects to architectures as discussed in [THUR97]. For example, a multimedia database system could use a commercial database system such as an object-oriented database system to manage multimedia objects. However, relationships between objects and the representation of temporal relationships may involve extensions to the database management system. That is, a DBMS together with an extension layer provide complete support to manage multimedia data. In the alternative case, both the extensions and the database management functions are integrated so that there is one database management system to manage multimedia objects as well as the relationships between the objects. These two types of architectures are illustrated in Figure 11-4 and Figure 11-5. Multimedia databases could also be distributed. In this case, we assume that each MM-DBMS is augmented with a Multimedia Distributed Processor (MDP) as illustrated in Figure 11-6.

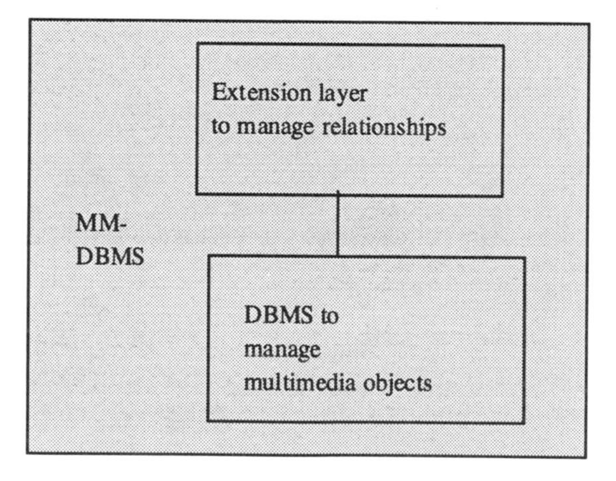

Figure 11-4. DBMS + Extension Layer

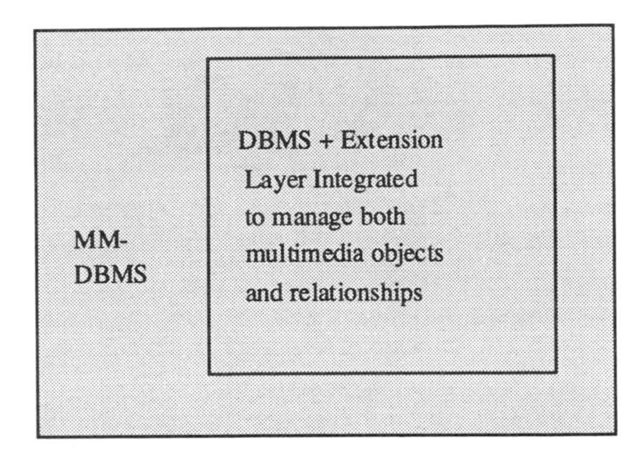

Figure 11-5. DBMS and Extension Layer Integrated

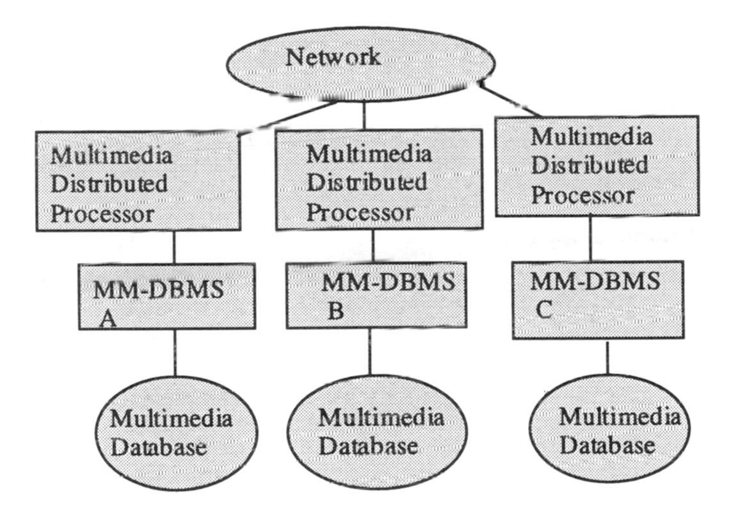

Figure 11-6. Distributed Multimedia DBMS

11.2.2 Data Modeling

In representing multimedia data, several features have to be supported. First of all, there has to be a way to capture the complex data types and all the relationships between the data. Various temporal constructs such as play-before, play-after, play-together, etc., have to be captured (see, for example, the discussion in [PRAB97]). Figure 11-7 illustrates a representation of a multimedia database. In this example, there are two objects: A and B. A consists of 2000 frames and B consists of 3000 frames. A consists of a time interval between 4/95 and 8/95 and B consists of a time interval between 5/95 and 10/95.

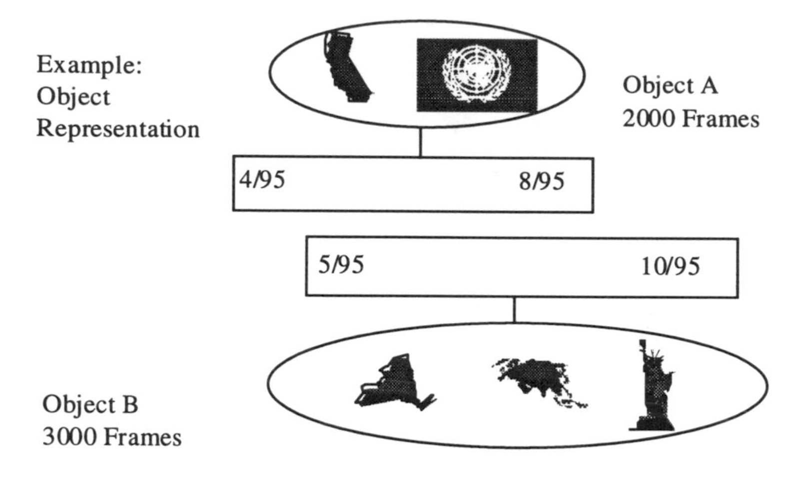

Figure 11-7. Data Representation

An appropriate data model is critical to represent an MM-DBMS. Relational, object-oriented, as well as object-relational data models have been examined to represent multimedia data. Some argue that relational models are better since they can capture relationships, while others argue that object models are better as they represent complex structures. In the example of Figure 11-7, with an object-oriented data model, each object in the figure would correspond to an object in the object model. The attributes of an object may be represented as instance variables and will include time interval, frames, and content description. With the relational model, the object would correspond to an instance of a relation. However with atomic values, it will be difficult to capture the attributes of the instance. In the case of the object-relational model, the attribute value of an instance could be an object. That is, for the instance that represents object A, the attribute value time interval would be the pair (4/95, 8/95). Representing object A with an object model is illustrated in Figure 11-8. Representing the same object with an object-relational model is illustrated in Figure 11-9. Note that one could build extensions to an existing data model to support complex relationships for multimedia data. These relationships may include temporal relationships between objects such as play together, play before, and play after.

Languages such as SQL are being extended for MM-DBMS (see, for example, [SQL3]). Others argue that object-oriented models are better as they can represent complex data types. It appears that both types of models have to be extended to capture the temporal constructs and other special features. Associated with a data model is a query language. The language should support the constructs needed to manipulate the multimedia database. For example, one may need to query to play frames 500 to 1000 of a video script.

In summary, several efforts are under way to develop appropriate data models for MM-DBMSs. Standards are also being developed. This is an area that has matured within the past couple of years.

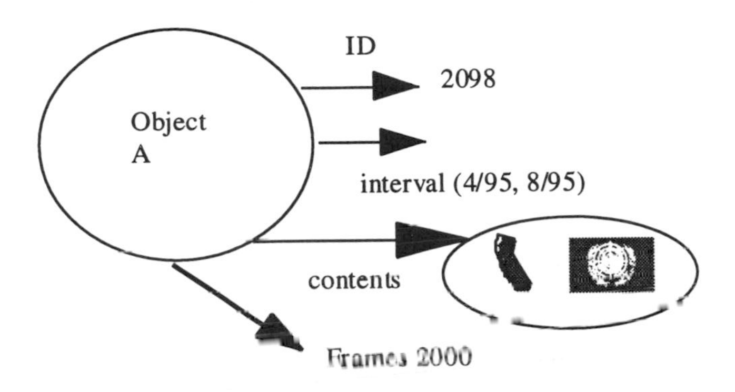

Figure 11-8. Data Representation with Object Model

ID	Interval	Contents	Frame
2098	(4/95, 8/95)		2000

Figure 11-9. Data Representation with Object-Relational Model

11.2.3 Functions of an MM-DBMS[16]

11.2.3.1 Overview

An MM-DBMS must support the basic DBMS functions. These include data manipulation, which includes query/update processing,

---

16 We discuss the functions briefly in this chapter. For more details and illustrations on MM-DBMS functions we refer to [THUR97]. A text devoted entirely to MM-DBMS is [PRAB97]. Other useful references are [NWOS96], [CHOR94], and [WOEL86].

transaction management, metadata management, storage management, and maintaining security and integrity. All of these functions are more complex since the data may be structured as well as unstructured. Furthermore, handling various data types such as audio and video is quite complex. In addition to these basic DBMS functions, an MM-DBMS must also support real-time processing to synchronize multimedia data types such as audio and video. Quality of service is an important aspect for MM-DBMS. For example, in certain cases, high quality resolution for images may not always be necessary. Special user interfaces are also needed to support different media.

This section provides an overview of the various functions. These include data manipulation such as query/update processing, browsing, and editing, transaction management, metadata management, data distribution, storage management, security, and integrity.

11.2.3.2 Data Manipulation

Data manipulation involves various aspects. Support for querying, browsing, and filtering the data is essential. Appropriate query languages are needed for this purpose. As discussed earlier, SQL extensions show much promise. In addition to just querying the data, one also may want to edit the data. That is, two objects may be merged to form a third object. One could project an object to form a smaller object. As an example, objects may be merged based on time intervals, and an object may be projected based on time intervals. Objects may also be updated in whole or in part. Much of the focus on MM-DBMS has been on data representation and data manipulation. Various algorithms have been proposed. Some of these algorithms have also been implemented in various systems [TKDE93].

11.2.3.3 Transaction Management

There has been some discussion as to whether transaction management is needed in MM-DBMS [ACM94]. We feel this is important, as in many cases annotations may be associated with multimedia objects. For example, if one updates an image, then its annotation must also be updated. Therefore, the two operations have to be carried out as part of a transaction. Unlike data representation and data manipulation, transaction management in an MM-DBMS is still a new area. Associated with transaction management are concurrency control and recovery. The issue is what are the transaction models? Are there special concurrency control and recovery mechanisms? Much research is needed in this area.

11.2.3.4 Metadata Management

Many of the metadata issues discussed for DBMSs also apply to MM-DBMSs. What is a model for metadata? What are the techniques for metadata management? In addition, there may be large quantities of metadata to describe, say, audio and video data. For example, in the case of video data, one may need to maintain information about the various frames. This information is usually stored in the metadata.

There are several other considerations. Metadata plays a crucial role in pattern matching. To do data analysis on multimedia data, one needs to have some idea as to what one is searching for. For example, in a video clip, if various images are to be recognized, then there must be some patterns already stored to facilitate pattern matching. Information about these patterns has to be stored in the metadata.

In summary, metadata management in an MM-DBMS is still a challenge. Some ideas were presented in [META96]. The emergence of Internet technologies makes this even more complex.

11.2.3.5 Storage Management

The major issues in storage management include developing special index methods and access strategies for multimedia data types. Content-based data access is important for many multimedia applications. However, efficient techniques for content-based data access are still a challenge. Other storage issues include caching data. How often should the data be cached? Are there any special considerations for multimedia data? Are there special algorithms? Also, storage techniques for integrating different data types are needed. For example, a multimedia database may contain video, audio, and text databases instead of just one data type. The display of these different data types has to be synchronized. Appropriate storage mechanisms are needed so that there is continuous display of the data.

Storage management for multimedia databases is also an area that has been given considerable attention. Several advances have been made during recent years [MDDS94].

11.2.3.6 Maintaining Data Integrity and Security

Maintaining data integrity will include support for data quality, integrity constraint processing, concurrency control, and recovery for multi-user updates, and accuracy of the data on output. The issues on integrity for database management systems in Chapter 2 are present for MM-DBMSs. However, enforcing integrity constraints remains a challenge. For example, what kinds of integrity constraints can be enforced on voice and video data? There is little research to address these issues.

Security mechanisms include supporting access rights and authorization. All of the security issues discussed in Chapter 2 also apply to MM-DBMSs. There are also additional concerns. For example, in the case of video data, should access control rules be enforced on entire scripts or frames? Again, little research has been done here.

11.2.3.7 Other Functions

Other functions for an MM-DBMS include quality of service processing, real-time processing, and user interface management. For example, with respect to quality of service, in some instances one may need continuous display of data, and in some instances one could tolerate breaks of service. One has to come up with appropriate primitives to specify quality of service. Real-time processing plays a major role since appropriate scheduling techniques are needed to display various types of media such as the voice with the video. Finally, appropriate multimodal interfaces are needed for inputting and displaying multimedia data.

11.3 MINING MULTIMEDIA DATA

11.3.1 Overview

Now that we have provided a brief overview of multimedia databases and discussed some of the essential concepts in terms of architectures, data models, and functions, in this section we will discuss the issues involved in mining and extracting information from these multimedia databases.

As stated earlier, multimedia data includes text, images, video, and audio. Text and images are still media, while audio and video are continuous media. The issues surrounding still and continuous media are somewhat different and have been explained in various texts and papers such as [PRAB97]. In this section we will consider text, image, video, and audio and consider how such data can be mined. First of all, what are the differences between mining multimedia data and topics such as text, image, and video retrieval? What is meant by mining such data? What are the developments and challenges?

Data mining has an impact on the functions of multimedia database systems discussed in the previous section. For example, the query processing strategies have to be adapted to handle mining queries if a tight integration between the data miner and the database system is the approach taken. This will then have an impact on the storage strategies.

Furthermore, the data model will also have an impact. At present, many of the mining tools work on relational databases. However, if object-relational databases are to be used for multimedia modeling, then data mining tools have to be developed to handle such databases.

Sections 11.3.2, 11.3.3, 11.3.4, and 11.3.5 will discuss text, image, video, and audio mining, respectively. In particular, the definition of mining, the developments, challenges, and directions are given. Then, in Section 11.3.6, we will briefly discuss the issues of mining combinations of data types.

11.3.2 Text Mining

Much of the information is now in textual form. This could be data on the web or library data or electronic books, among others. One of the problems with text data is that it is not structured as relational data. In many cases it is unstructured and in some cases it is semistructured. Semistructured data, for example, is an article that has a title, author, abstract, and paragraphs. The paragraphs are not structured, while the format is structured.

Information retrieval systems and text processing systems have been developed for more than a few decades. Some of these systems are quite sophisticated and can retrieve documents by specifying attributes or key words. There are also text processing systems that can retrieve associations between documents. So we are often asked what the difference between information retrieval systems and text mining systems is?

We define text mining to be data mining on text data. Text mining is all about extracting patterns and associations previously unknown from large text databases. The difference between text mining and information retrieval is analogous to the difference between data mining and database management. There is really no clear difference. Some of the recent information retrieval and text processing systems do discover associations between words and paragraphs, and therefore can be regarded as text mining systems.

Next, let us examine the approaches to text mining. Note that many of the current tools and techniques for data mining work for relational databases. Even for data in object-oriented databases, rarely do we hear about data mining tools for such data. Therefore, current data mining tools cannot be directly applied to text data. Some of the current directions in mining unstructured data include the following.

- Extract data and/or metadata from the unstructured databases possibly by using tagging techniques, store the extracted data in structured databases, and apply data mining tools on the structured databases. This is illustrated in Figure 11-10.

- Integrate data mining techniques with information retrieval tools so that appropriate data mining tools can be developed for unstructured databases. This is illustrated in Figure 11-11.

- Develop data mining tools to operate directly on unstructured databases. This is illustrated in Figure 11-12.

Now, while converting text data into relational databases, one has to be careful so that there is no loss of key information. As we have stated before, unless you have good data you cannot mine the data effectively and expect to get useful results. One needs to create a sort of a warehouse first before mining the converted database. This warehouse is essentially a relational database that has the essential data from the text data. In other words, one needs a transformer that takes a text corpus as input and outputs tables that have, for example, the keywords from the text.

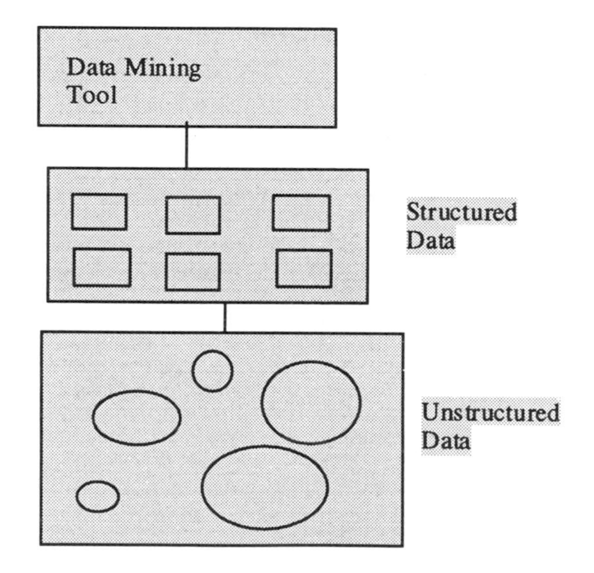

Figure 11-10. Converting Unstructured Data to Structured Data for Mining

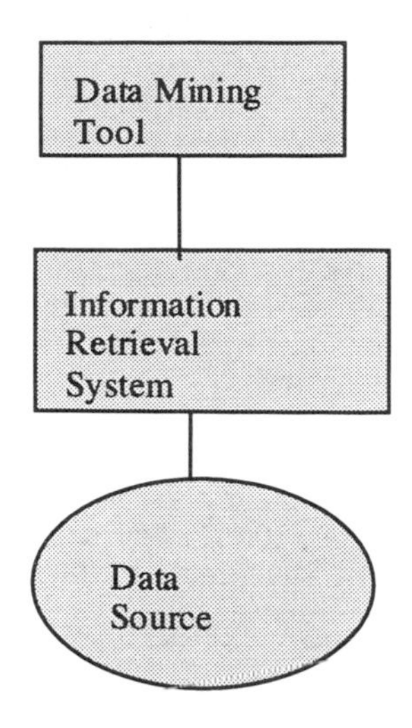

Figure 11-11. Augmenting an Information Retrieval System

As an example, in a text database that has several journal articles, one could create a warehouse with tables containing the following attributes: author, date, publisher, title, and keywords. From the keywords, one can form associations. The keywords in one article could be "Belgium, nuclear weapons" and the keywords in another article could be "Spain, nuclear weapons." The data miner could make the association that authors from Belgium and Spain write articles on nuclear weapons.

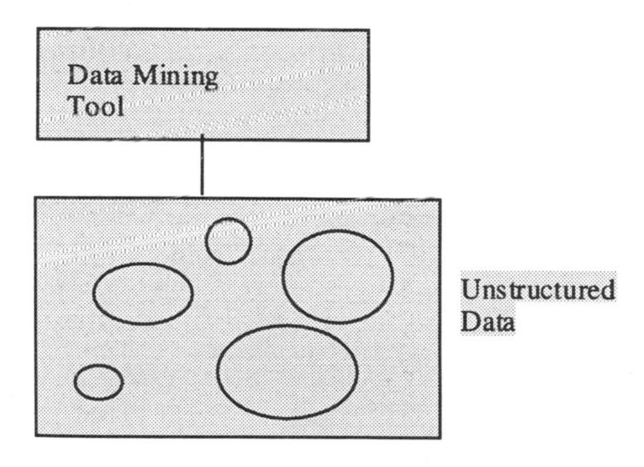

Figure 11-12. Mining Directly on Unstructured Data

Note that we are only in the beginning of text mining. In the longer-term approach we would want to develop tools directly to mine text data. These tools have to read the text, understand the text, put out pertinent information about the text, and then make associations between different documents. We are far from developing such sophisticated text mining tools. However, the work reported by Tsur, Ullman, and Clifton et al. (see, for example, [TSUR98]) is the first step in the right direction toward text mining. Some interesting early work on text mining was reported in [FELD95].

11.3.3 Image Mining

If text mining is still in the early research stages, image mining is an even more immature technology. In this section, we will examine this area and discuss the current status and challenges.

Image processing has been around for quite a while. We have image processing applications in various domains including medical imaging for cancer detection, processing satellite images for space and intelligence applications, and also handling hyperspectral images. Images include maps, geological structures, biological structures, and many other entities. Image processing has dealt with areas such as detecting abnormal patterns which deviate from the norm, retrieving images by content, and pattern matching.

The main question here is what is image mining? How does it differ from image processing? Again we do not have clear cut answers. One can say that while image processing is focusing on detecting abnormal patterns as well as retrieving images, image mining is all about finding unusual patterns. Therefore, one can say that image mining deals with making associations between different images from large image databases

Clifton et al. [CLIF98a] have begun work in image mining. Initially their plan was to extract metadata from images and then carry out mining on the metadata. This would essentially be mining the metadata in relational databases. However, after some consideration it was felt that images could be mined directly. The challenge then is to determine what type of mining outcome is most suitable. One could mine for associations between images, cluster images, classify images, as well as detect unusual patterns. One area of research being pursued by Clifton et al. is to mine images and find out whether there is anything unusual. So the approach is to develop templates that generate several rules about the images, and from there, apply the data mining tools to see if unusual patterns can be obtained. However, the mining tools will not tell us why these patterns are unusual. Figure 11-13 shows an image with some unusual patterns.

Note that detecting unusual patterns is not the only outcome of image mining. However, this is just the beginning. We need to conduct more research on image mining to see if data mining techniques could be used to classify, cluster, and associate images. Image mining is an area with applications in numerous domains including space images, medical images, and geological images.

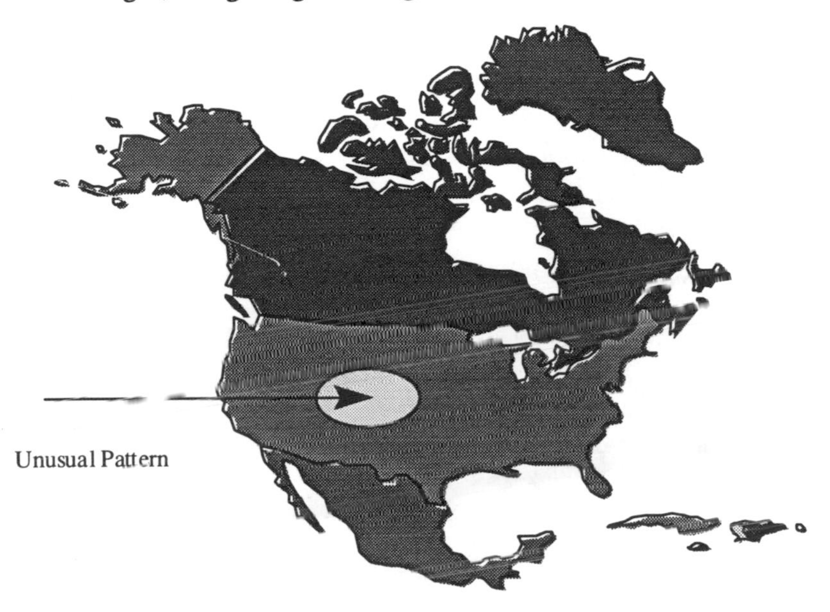

Figure 11-13. Image Mining

11.3.4 Video Mining

Mining video data is even more complicated than mining image data. One can regard video to be a collection of moving images, much like animation. Video data management has been the subject of much research. The important areas include developing query and retrieval techniques for video databases, including video indexing, query languages, and optimization strategies. The first question one asks yet again is what is the difference between video information retrieval and video mining? Unlike image and text mining, we do not have any clear idea of what is meant by video mining. For example, one could examine video clips and find associations between different clips. Another example would be to find unusual patterns in video clips. But how is this different from finding unusual patterns in images? So, the first step to successful video mining is to have a good handle on image mining.

Let us examine pattern matching in video databases. Should one have predefined images and then match these images with the video data? Is there any way one can do pattern recognition in video data by specifying what one is looking for and then try to do feature extraction for the video data? If this is video information retrieval what then is mining video data? To be consistent with our terminology we can say that finding correlations and patterns previously unknown from large video databases is video mining. So by analyzing a video clip or multiple video clips, one comes to conclusions about some unusual behavior. People in the video who are unlikely to be there, yet have occurred two or three times could mean something significant. Another way to look at the problem is to capture the text in video format and try and make the associations one would carry out with text but this time use the video data instead.

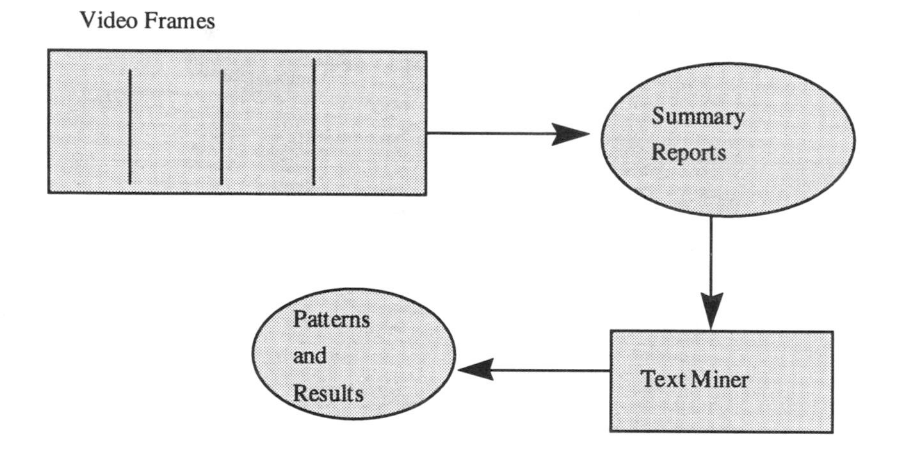

Figure 11-14. Mining Text Extracted from Video

Unlike text and image mining where our ideas have been less vague, the discussion here on video mining is quite preliminary. This is mainly because there is so little known on video mining. Even the word video mining is something very new, and to date we do not have any concrete results reported on this. We do have a lot of information on analyzing video data and producing summaries. Now one could mine these summaries, which would amount to mining text as shown in Figure 11-14. One good example of this effort is the work by Merlino et al. on summarizing video news [MERL97]. Converting the video mining problem to a text mining problem is reasonably well understood. However, the challenge is mining video data directly, and more importantly, knowing what we want to mine. With the emergence of the web, video mining becomes even more important. An example of direct video mining is illustrated in Figure 11-15.

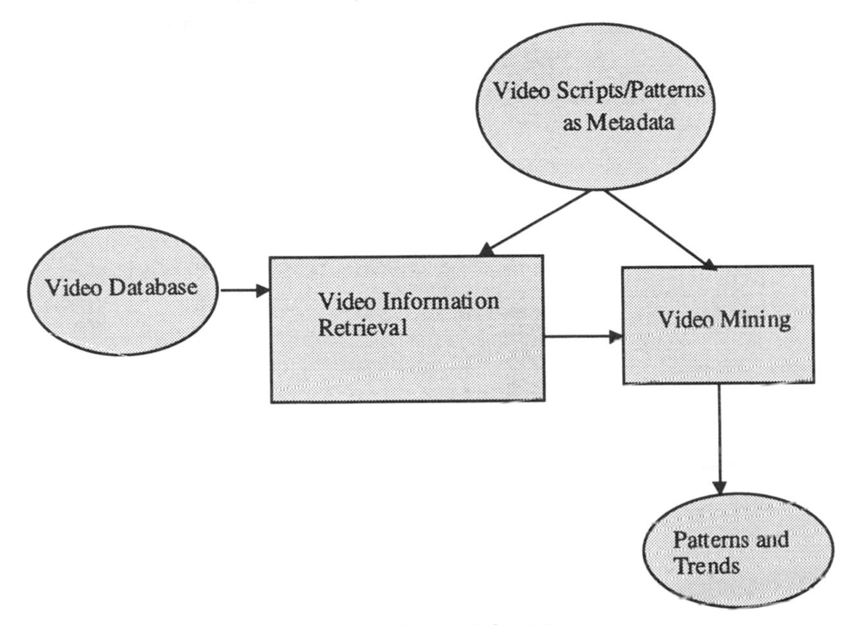

Figure 11-15. Direct Video Mining

11.3.5 Audio Mining

Since audio is a continuous media type like video, the techniques for audio information processing and mining are similar to video information retrieval and mining. Audio data could be in the form of radio, speech, or spoken language. Even television news has audio data, and in this case audio may have to be integrated with video and possibly text to capture the annotations and captions.

To mine audio data, one could convert it into text using speech transcription techniques and other techniques such as keyword extraction and then mine the text data as illustrated in Figure 11-16. On the other hand, audio data could also be mined directly by using audio information processing techniques and then mining selected audio data. This is illustrated in Figure 11-17.

In general, audio mining is even more primitive than video mining. While a few papers have appeared on text mining and even fewer on image and video mining, work on audio mining is just beginning.

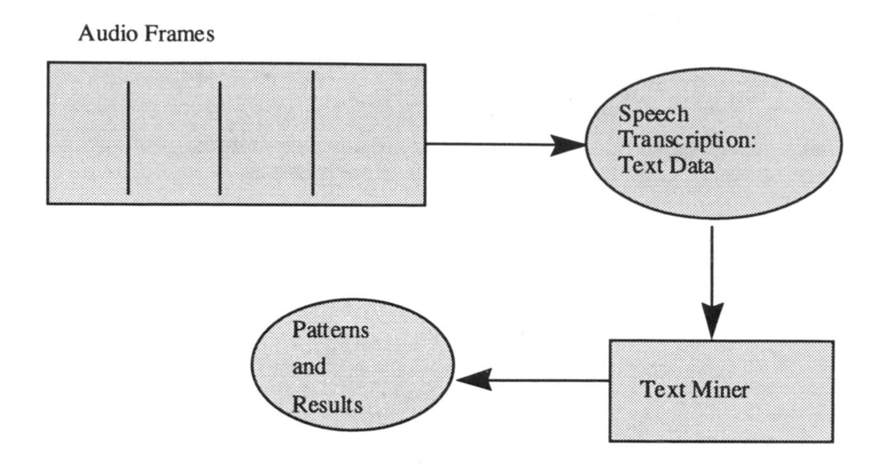

Figure 11-16. Mining Text Extracted from Audio

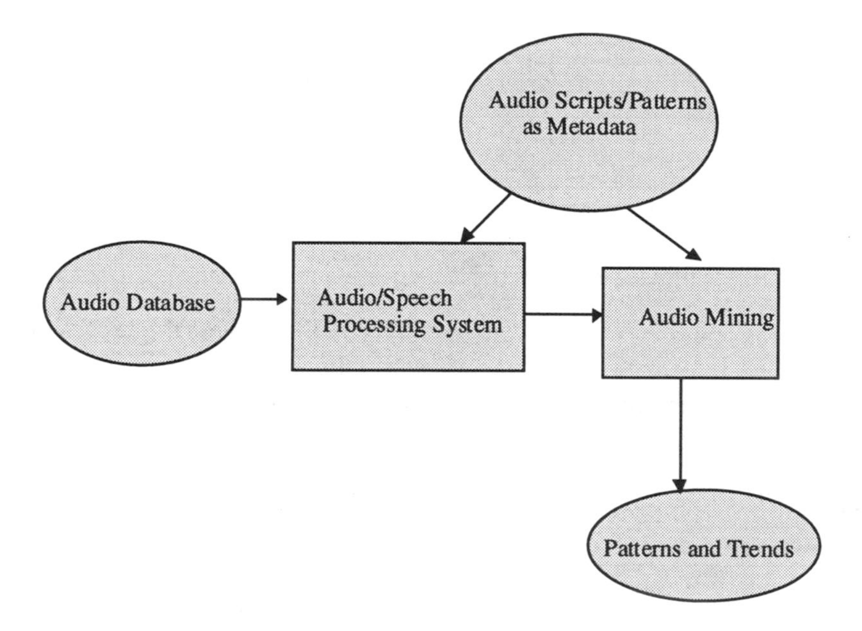

Figure 11-17. Direct Audio Mining

11.3.6 Mining Combinations of Data Types

The previous section discussed mining on individual data types such as text, images, video, and audio. If we are to mine multimedia data, then we need to mine combinations of two or more data types such as text and images, text and video, or text, audio, and video. In this section we will briefly discuss some of the issues on multimedia data mining.

Handling combinations of data types is very much like dealing with heterogeneous databases. For example, each database in the heterogeneous environment could contain data belonging to multiple data types. These heterogeneous databases could be first integrated and then mined or one could apply mining tools on the individual databases and then combine the results of the various data miners. These two scenarios are illustrated in Figures 11-18 and 11-19. In both cases the Multimedia Distributed Processor (MDP) plays a role. If the data is to be integrated before being mined, then this integration is carried out via the MDPs. If the data is to be mined first, the data miner augments the corresponding MM-DBMS and the results of the data miners are integrated via the MDPs

Since there is much to be done on mining individual data types such as text, images, video, and audio, mining combinations of data types is still a challenge. Once we have a better handle on mining individual data types, we can then focus on mining combinations of data types.

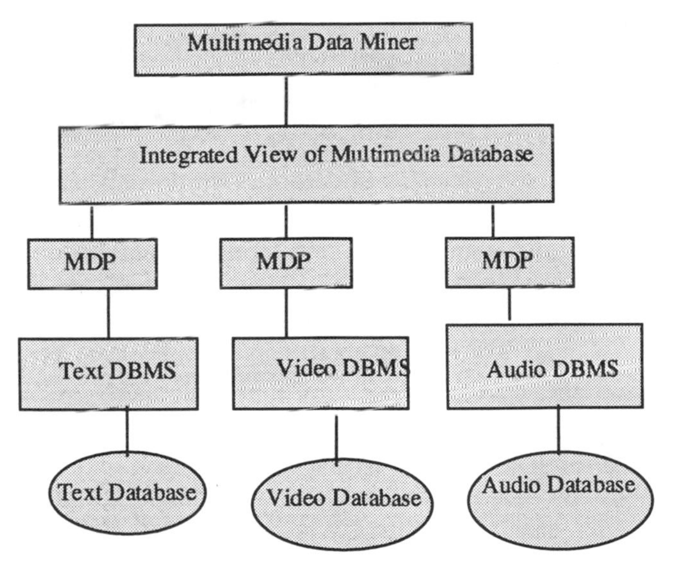

Figure 11-18. Integration and Then Mining

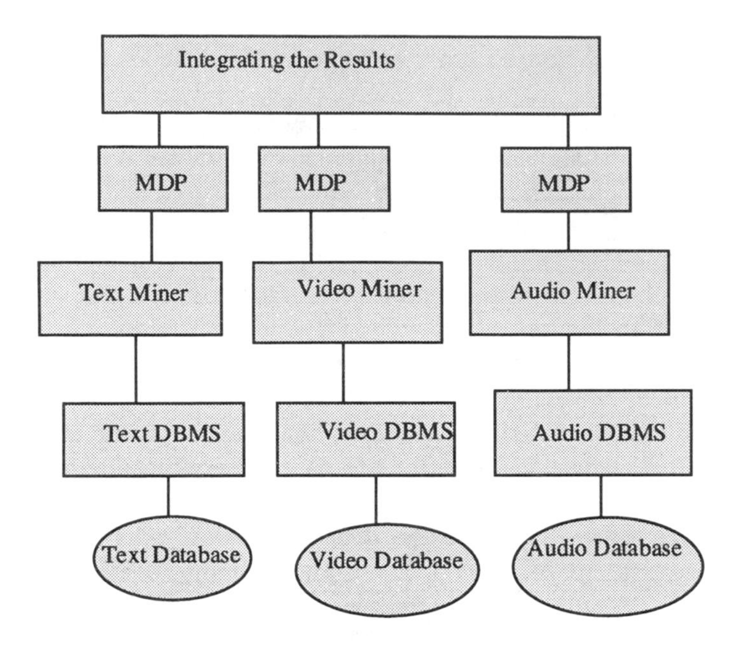

Figure 11-19. Integration and Then Mining

11.4 SUMMARY

In this chapter we started with a definition of multimedia database management systems and then provided an overview of these systems. In particular, different types of architectures, data models, and functions of these systems were discussed. Then we addressed data mining for multimedia data. We focused on four types of media: text, image, video, and audio data. We defined what data mining meant for such data and discussed the developments and the challenges. We also provided directions for mining such data. Finally, we briefly discussed issues on mining combinations of data types.

We have mainly addressed mining individual data types. Mining multimedia data would involve addressing a combination of two or more media types. As we learn more about mining text, images, video, and audio, we can expect progress to be made on multimedia data mining. Mining combinations of data such as video and text, video, audio, and text, or image and text remains a challenge.

We believe that as progress is made on multimedia data management and data mining, we will begin to see tools emerge on mining multimedia data. At present, data mining tools work largely on relational databases. However, in the future we can expect to see multimedia data mining tools as well as tools for mining object databases. Research still needs to be done in this area.

CHAPTER 12

DATA MINING AND THE WORLD WIDE WEB

12.1 OVERVIEW

A key emerging technology is digital libraries and Internet database management.[17] Data on the Internet, also known as the world wide web (WWW), has to be effectively managed. Consequently, this data has to be mined to obtain patterns and trends. This chapter describes some of the issues on Internet data management and mining. This topic is also called web data mining.

Digital libraries are essentially digitized information distributed across several sites. The goal is for users to access this information in a transparent manner. The information could contain multimedia data such as voice, text, video, and images. The information could also be stored in structured databases such as relational and object-oriented databases.

The explosion of the users on the Internet and the increasing number of world wide web servers are rapidly advancing digital libraries. This is because digital libraries are usually hosted on networks including the Internet. That is, users can access the various digital libraries across the Internet. There is no single technology for digital libraries. It is a combination of many technologies including heterogeneous database management, mass storage management, collaborative/workflow computing, multimedia database management, intelligent agents and mediators, and data mining. For example, the heterogeneous information sources have to be integrated so that users access the servers in a transparent and timely manner. Security and privacy is becoming a major concern for digital libraries. So are other issues such as copyright protection and ownership of the data. Policies and procedures have to be set up to address these issues.

Major national initiatives are under way to develop digital library technologies. The agencies funding digital library work include the National Science Foundation, Defense Advanced Research Projects Agency, and the National Aeronautical and Space Administration. In addition, there are numerous projects funded by organizations such as

---

[17] Note that we have used the term digital libraries and Internet database management interchangeably. Many of the issues for digital libraries are present for Internet database management. The Internet began as a research effort funded by the U.S. Government. It is now the most widely used network in the world.

the Library of Congress to develop digital library technologies (see, for example, [ACM95]). Various conferences and workshops have also been established recently devoted entirely to digital libraries (see, for example, [DIGI95]).

With information overload on the web, it is highly desirable to mine the data and extract patterns and relevant information to the user. This will make the task of browsing on the Internet so much easier for the user. Therefore, there has been a lot of interest on mining the web, which is also now called web mining. This is essentially mining the databases on the web or mining the usage patterns so that helpful information can be provided to the user. This chapter is devoted to a discussion of some of the preliminary work on web mining.

The organization of this chapter is as follows. We first provide an overview of digital libraries and Internet database management in Section 12.2. In particular, some of the technology integration issues for digital libraries, potential uses with digital libraries, and functions of digital libraries are described. Data mining on the web is the subject of Section 12.3. The chapter is summarized in Section 12.4.

12.2 INTERNET DATABASE MANAGEMENT AND DIGITAL LIBRARIES

12.2.1 Technologies

Various technologies have to be integrated to develop digital libraries. Some of the important data management technologies for digital libraries are data mining, multimedia database management, and heterogeneous database integration. In addition, some of the supporting technologies such as agents, distributed object management, and mass storage are also important. Figure 12-1 illustrates the various digital library technologies.

Integration of these technologies is a major challenge. First of all, appropriate Internet access protocols have to be developed. In addition, interface definition languages play a major role in the interoperability of different systems. Due to the large amount of data, integration of mass storage with data management will be critical. Data mining is needed to extract information from the databases. Multimedia technology combined with hypermedia technology is necessary for browsing multimedia data. Distributed object management will play a major role, especially since the number of data sources to be integrated may be large. The remaining subsections describe the uses of digital libraries, architectural aspects, and functions of digital libraries.

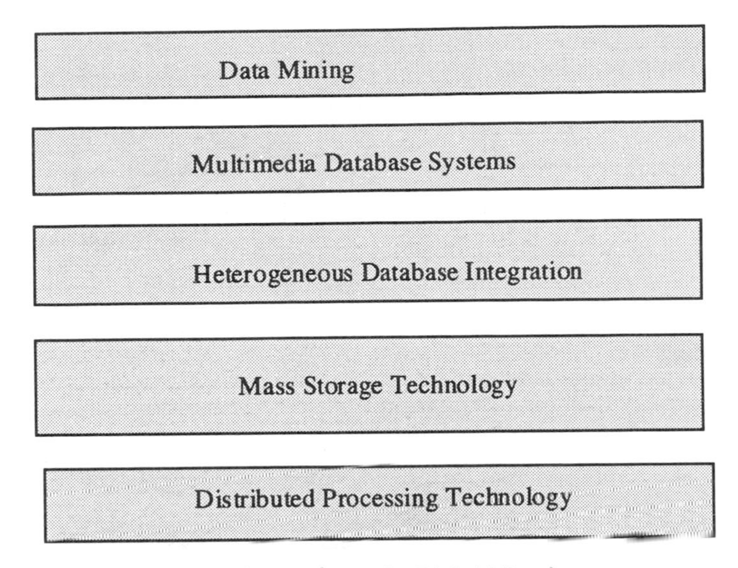

Figure 12-1. Some Technologies for Digital Libraries

12.2.2 Uses of Digital Library

An example of a digital library is illustrated in Figure 12-2. The idea here is that there are a certain number of sites participating in this library. Note that in theory the library could also have an unlimited number of users. However, many organizations want to share the data between a certain number of groups.

The information in the form of servers, databases, and tools belongs to the library. The participating sites could place this information or it could be placed by someone who is designated to maintain the library. Users then query and access the information in the library.

Figure 12-2 also illustrates the use of agents to maintain the library. These agents locate resources for users, maintain the resources, and even filter out information so that users only get the information they want. Agents are essentially intelligent processes. They may communicate with each other in carrying out a specific task. Agents may carry out a number of other functions including query processing. For a discussion of intelligent agents we refer to [MATT98].

One can also take advantage of the digital library technology for collaborative work environments. Suppose organization A wants to develop some technology such as integrating heterogeneous databases. They access the WWW and find out the names of other organizations who already have developed such systems. They may like what is said about the system developed by organization B. They contact organization B and get a demonstration of the system through the Internet.

Collaborative work environments are discussed in more detail in [THUR97].

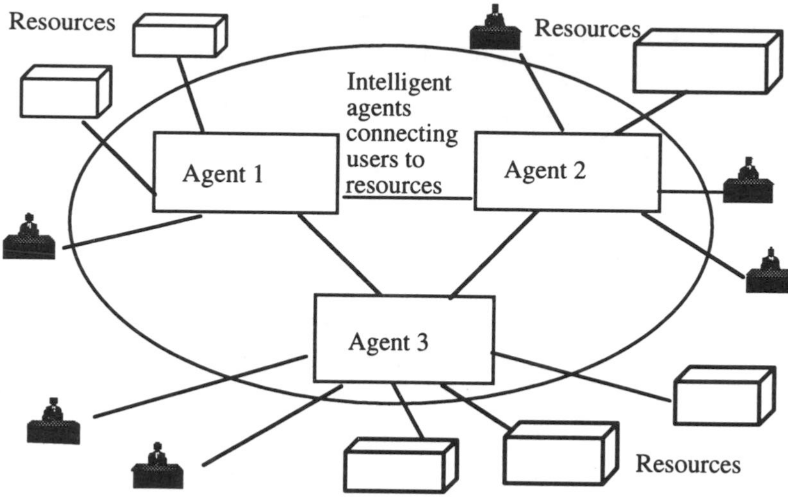

Figure 12-2. Example Digital Library

12.2.3 Architectural Aspects

Data management has become an integral part of digital libraries. Database management system vendors are now building interfaces to the Internet. Query languages like SQL are embedded into Internet access languages. For example, as shown in Figure 12-3, DBMS vendors A and B could make their data available to applications C and D. DBMS vendors are also developing an interface to the Java programming environment.[18] Essentially, what this all means is that heterogeneous databases are integrated through the Internet.

Architectural aspects also include the various ways one can connect the web databases and servers to the clients. Details are given in [IEEE98]. In one approach, a relational interface is provided to integrate the various web databases and documents. This work was carried out by Stanford University and is called the Junglee system and is now being commercialized (see, for example, [STAN98] and [JUNG98]). In another, the data in the relational database is transformed into web documents by some gateways and web servers manage the web documents. Web clients and servers communicate with each other through various Internet protocols. This approach is illustrated in Figure 12-4.

---

[18] Java is a product of JavaSoft, a subsidiary of Sun Microsystems.

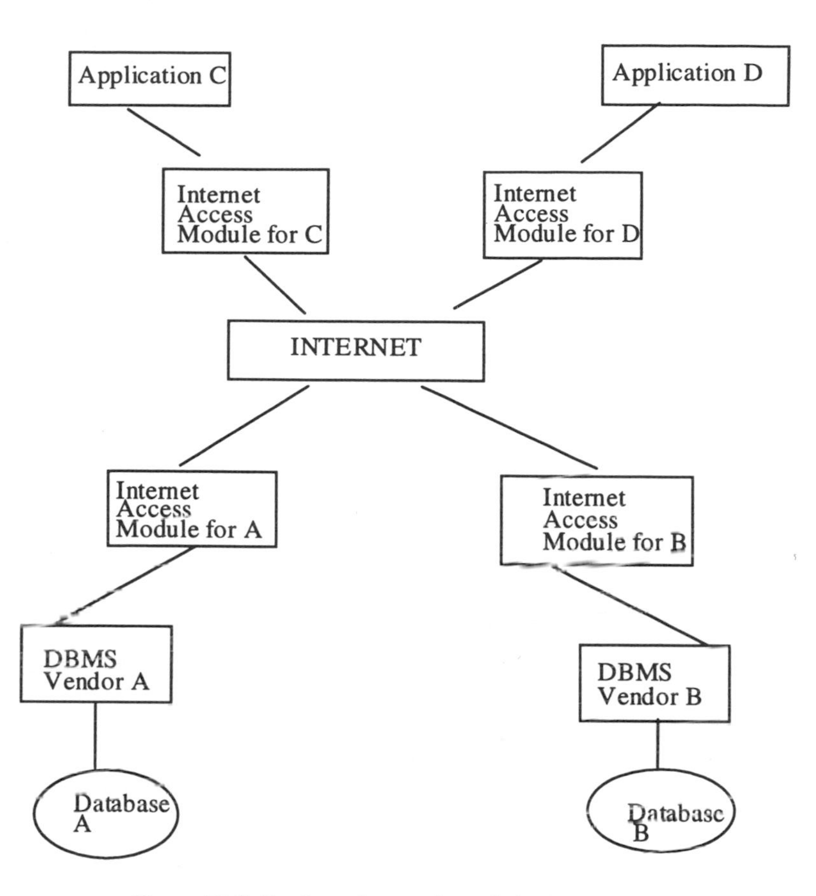

Figure 12-3. Database Access through the Internet

12.2.4 Database Management Functions

Database management functions for digital libraries include those such as query processing, metadata management, security, and integrity as illustrated in Figure 12-5. In [THUR96b] we have examined various database management system functions and discussed the impact of Internet database access on these functions. Some of the issues are discussed here.

One of the major functions is data representation. The question is, is there a need for a standard data model for digital libraries and Internet database access? Is it possible to develop such a standard? If so, what are the relationships between the standard model and the individual models used by the database? Note that various data representation schemes such as SGML (Generalized Markup Language), HTML (Hypertext Markup Language), XML (Extended Markup Language), and ODA (Office Document Architecture) are among the schemes being

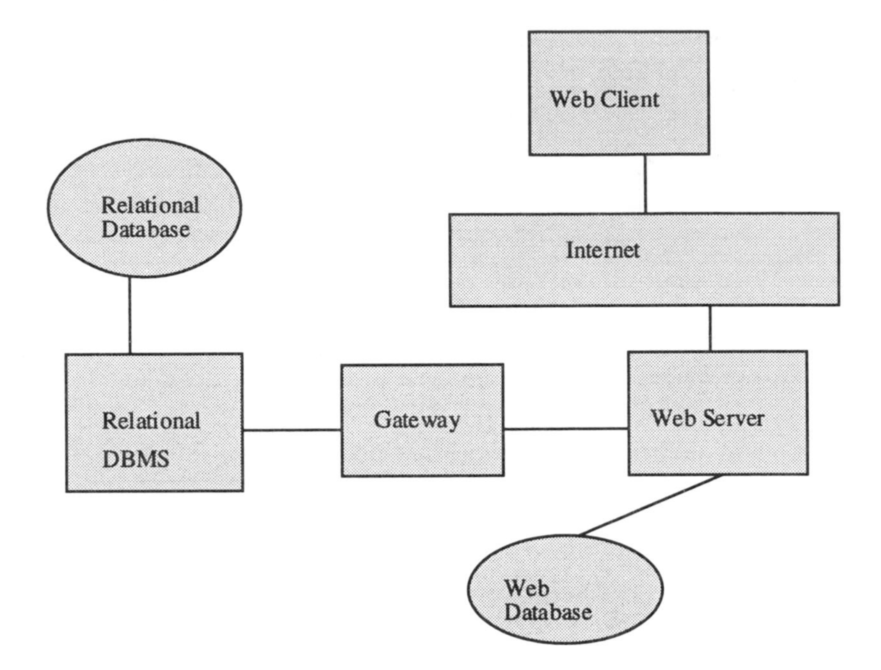

Figure 12-4. Database Access through Gateways

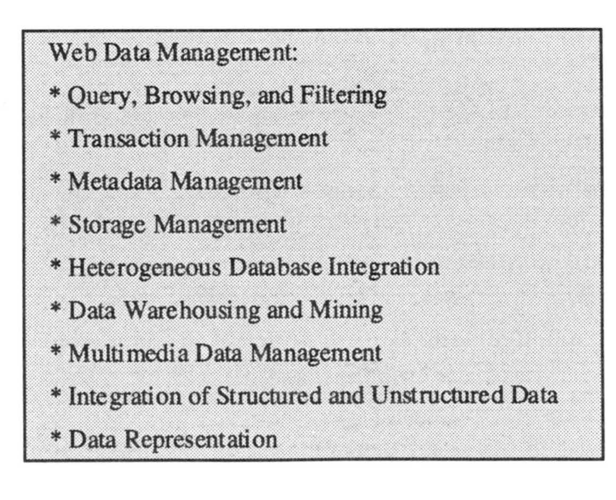

Figure 12-5. Web Data Management Functions

examined (see, for example, [ACM96b]).[19] Are they sufficient, or is another representation scheme needed?

Querying and browsing are two of the key functions. First of all, an appropriate query language is needed. Since SQL is a popular language, appropriate extensions to SQL may be desired. Query processing involves developing a cost model. Are there special cost models for Internet database management? With respect to browsing operation, the query processing techniques have to be integrated with techniques for following links. That is, hypermedia technology has to be integrated with database management technology.

Updating digital libraries could mean different things. One could create a new web site, place servers at that site, and update the data managed by the servers. The question is, can a user of the library send information to update the data at a web site? The issue here is security privileges. If the user has write privileges, then he could update the databases that he is authorized to modify. Agents and mediators could be used to locate the databases as well as to process the update.

Transaction management is essential for many applications. There may be new kinds of transactions on the Internet. For example, various items may be sold through the Internet. In this case, the item should not be locked immediately when a potential buyer makes a bid. It has to be left open until several bids are received and the item is sold. Therefore, special transaction models are needed. Appropriate concurrency control and recovery techniques have to be developed for the transaction models.

Metadata management is a major concern for digital libraries. The question is, what is metadata? Metadata describes all of the information pertaining to the library. This could include the various web sites, the types of users, access control issues, and policies enforced. Where should the metadata be located? Should each participating site maintain its own metadata? Should the metadata be replicated or should there be a centralized metadata repository? Metadata in such an environment could be very dynamic especially since the users and the web sites may be changing continuously.

Managing multimedia data is a concern. All of the issues described in Chapter 11 apply for digital libraries. In addition, synchronization and distribution issues are more complex. Data from different web servers may have to be synchronized before being displayed to the user.

---

[19] Note that XML is becoming an extremely popular technology for document interchange on the web. Some predict that the future of web data management will be based on XML. For information on XML we refer to [XML98].

Storage management for Internet database access is a complex function. Appropriate index strategies and access methods for handling multimedia data are needed. In addition, due to the large volumes of data, techniques for integrating database management technology with mass storage technology are also needed.

Security and privacy is a major challenge. Once you put the data at a site, who owns the data? If a user copies the data from a site, can he distribute the data? Can he use the information in papers that he is writing? Who owns the copyright to the original data? What role do digital signatures play? Mechanisms for copyright protection and plagiarism detection are needed. In addition, issues such as handling heterogeneous security policies will be of concern.[20]

Maintaining the integrity of the data is critical. Since the data may originate from multiple sources around the world, it will be difficult to keep tabs on the accuracy of the data. Data quality maintenance techniques need to be developed for digital libraries and Internet database access. For example, special tagging mechanisms may be needed to determine the quality of the data.

Heterogeneous database access, data warehousing, and data mining are important functions of digital libraries. The various heterogeneous data sources have to be integrated to provide transparent access to the user, and some of the details were discussed in Chapter 10. In some cases, the data sources have to be integrated into a warehouse. Data mining helps the users to extract meaningful information from the numerous data sources. Since the data in the libraries could have different semantics and syntax, it will be difficult to extract useful information. Sophisticated data mining tools are needed for this purpose. A discussion on data mining and its impact on the world wide web is the subject of the next section.

Interoperability between heterogeneous data sources is a major issue for digital libraries. Modules for interconnecting the different data sources and handling various types of heterogeneity are needed. Due to the potentially large number of data sources, one could expect distributed object management technology to play a major role for digital libraries and database access through the Internet. For example, different data sources may be encapsulated as objects, and they may commu-

---

20 Also, there has been a lot of discussion on the notion of a "firewall" to protect the internal information from external users. We do not address firewall issues in this chapter. For more details we refer the reader to various papers on Internet security such as the ones given in [NISS97].

nicate through some object request broker. Details are discussed in [THUR97].

12.3 WEB DATA MINING

Mining the data on the web is one of the major challenges faced by the data management and mining community as well as those working on web information management and machine learning. There is so much data and information on the web that extracting the useful and relevant information for the user is the real challenge here. When one scans through the web it becomes quite daunting, and soon we get overloaded with data. The question is, how do you convert this data into

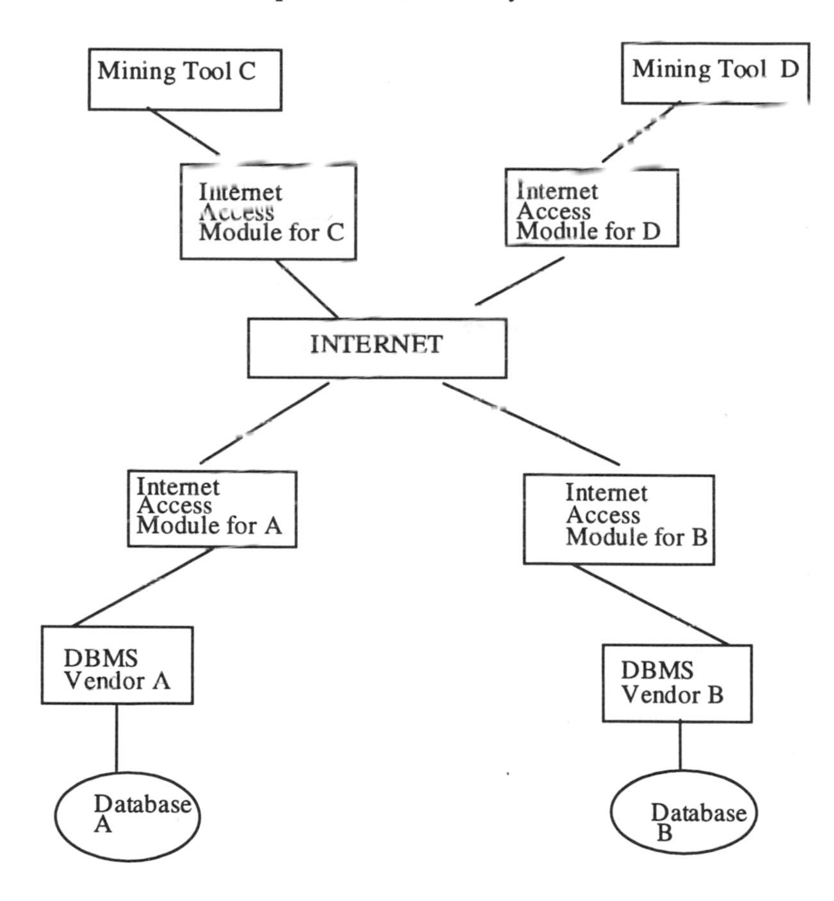

Figure 12-6. Data Mining through the Web

information and subsequently knowledge so that the user only gets what he wants? Furthermore, what are the ways of extracting information previously unknown from the data on the web? In this section we discuss various aspects to web mining.

One simple solution is to integrate the data mining tools with the data on the web. This is illustrated in Figure 12-6. This approach works well especially if the data is in relational databases. Therefore, one needs to mine the data in the relational databases with the data mining tools that are available. These data mining tools have to develop interfaces to the web. For example, if a relational interface is provided as in the Junglee system (see, for example, [JUNG98]), then SQL-based mining tools could be applied to the virtual relational database as illustrated in Figure 12-7.

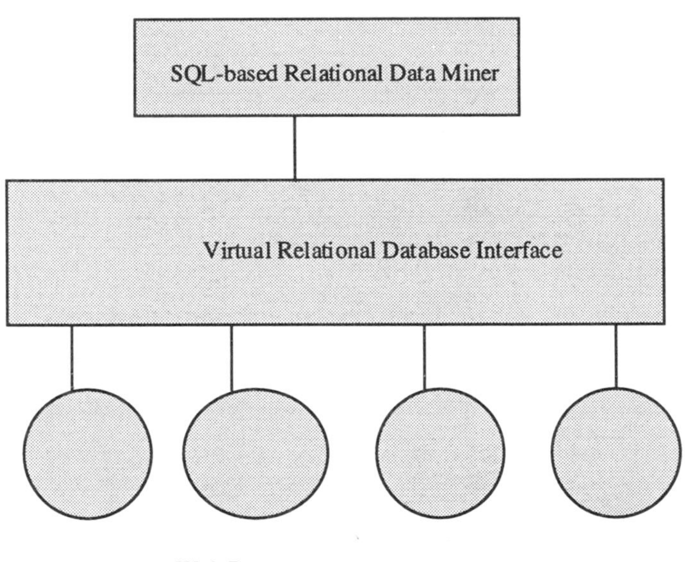

Figure 12-7. Web Mining on Virtual Relational Databases

Unfortunately the web world is not so straightforward. Much of the data is unstructured and semistructured. There is a lot of imagery data and video data. Providing a relational interface to all such databases may be complicated. The question is, how do you mine such data? In Chapter 11 we discussed various aspects to mining multimedia data. In particular, we focused on mining text, images, video, and audio data. One needs to develop tools first to mine multimedia data and then we can focus on developing tools to mine such data on the web. We illustrate a scenario for multimedia mining on the web in Figure 12-8 where multimedia databases are first integrated and then mined. Much of the previous discussion has focused on integrating data mining tools with the databases on the web. In many cases the data on the web is not in databases. It is on various servers. Therefore, the challenge is to organize the data on these servers. Some form of data warehousing technology may be needed here to organize the data to be mined. A scenario is illustrated in Figure 12-9. There is little work on developing some sort of data warehousing technology for the web to facilitate mining.

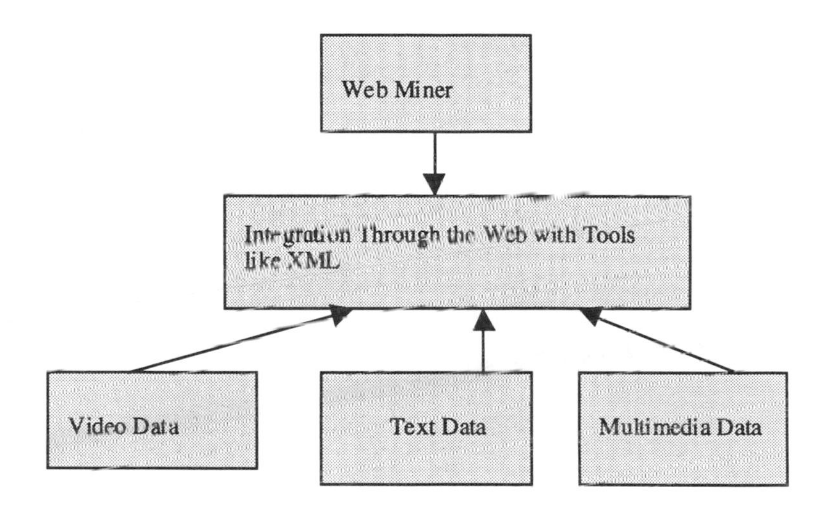

Figure 12-8. Multimedia Web Mining

Another area that really needs attention is visualization of the data on the web [THUR96c]. Much of the data is unorganized and difficult for the user to understand. Furthermore, as discussed in Chapter 4, mining is greatly facilitated by visualization. Therefore, developing appropriate visualization tools for the web will greatly facilitate mining the data. These visualization tools could aid in the mining process as illustrated in Figure 12-10.

Another aspect to mining on the web is to collect various statistics and determine which web pages are likely to be accessed based on various usage patterns. Research in this direction is being conducted by various groups including by Morey et al. [MORE98b]. Here, based on usage patterns of various users, trends and predictions are made as to the likely web pages a user may want to scan. Therefore, based on this information, a user can have guidance as to the web pages he may want

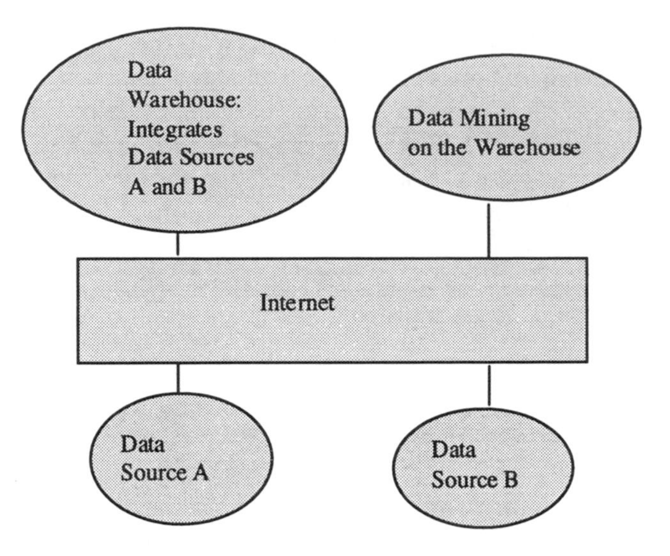

Figure 12-9. Data Warehousing and Mining on the Internet

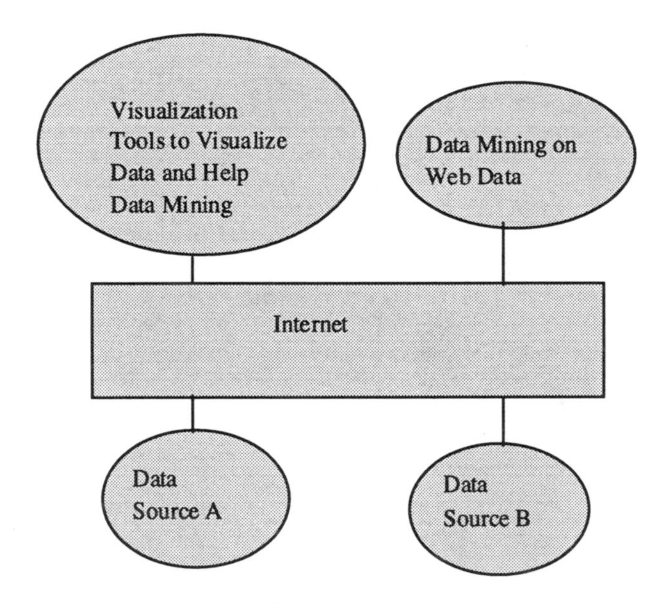

Figure 12-10. Data Mining and Visualization on the Web

to browse, as illustrated in Figure 12-11. This will facilitate the work a user has to do with respect to scanning various web pages. Note that while the previous paragraphs in this section focused on developing data mining tools to mine the data on the web, here we are focusing on using mining to help with the web browsing process. We can expect to see many results in this area.

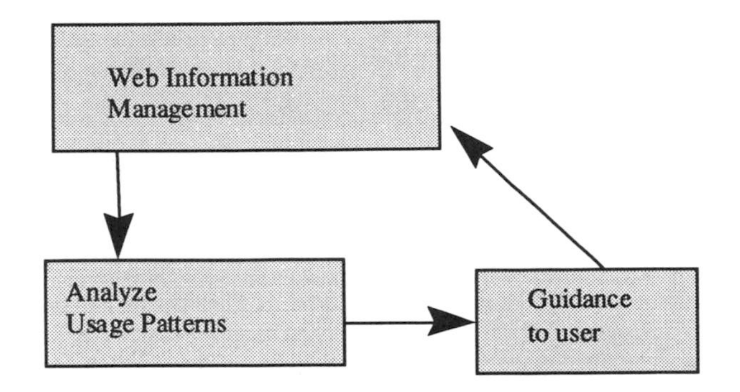

Figure 12-11. Analyzing Usage Patterns and Predicting Trends

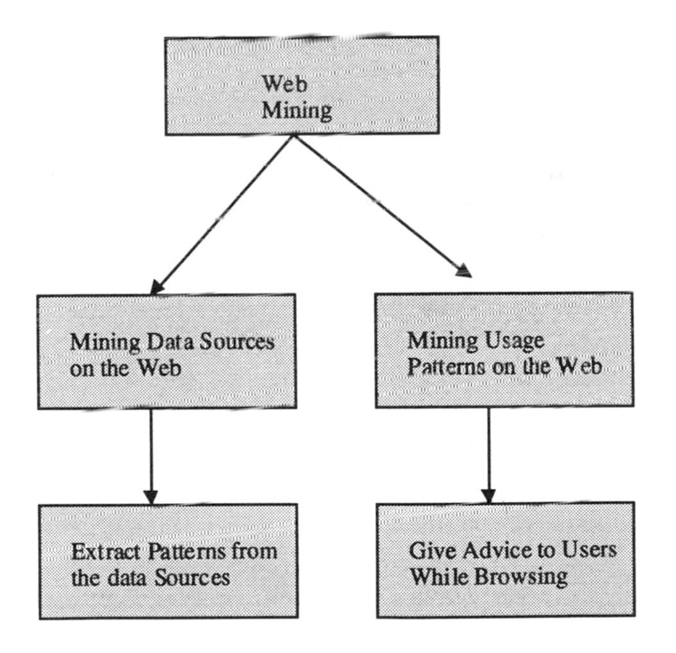

Figure 12-12. Taxonomy for Web Mining

Mining can also be used to give only selective information to the user. For example, many of us are flooded with email messages daily. Some of these messages are not relevant for our work. One can develop tools to discard the messages that are not relevant. These tools could be simple filtering tools or sophisticated data mining tools. Similarly, these data mining tools could also be used to display only the web pages in which a user is interested.

Cooley et al. [COOL98] have specified a taxonomy for web mining. They divide web mining into two categories. One is to mine and get patterns from the web data. The other is to mine the URLs and other web links to help the user with various activities on the web. Figure 12-12 illustrates this taxonomy.

Recently, various standards have been developed by organizations such as ISO (International Standards organization), W3C (World Wide Web Consortium), and OMG (Object Management Group) for Internet data access and management. These include models, specification languages, and architectures. One of the developments is XML (Extended Markup Language) for writing what is called a Document Type Definition that allows the document to be interpreted by the person receiving the document (see, for example, [XML98]). Connection to data mining and standards such as XML are largely unexplored. However, one could expect data mining languages to be developed for the web.

In summary, several technologies have to work together to effectively mine the data on the web. These include data mining on multimedia data, mining tools to predict trends and activities on the web, as well as technologies for data management on the web, data warehousing, and visualization. There is active research in web mining and we can expect to see much progress to be made here.

12.4 SUMMARY

This chapter has discussed the emerging topic of web data mining. First we provided some background on web data management. We have also called this Internet database management or digital libraries. In particular, the definition of a digital library, operation of a digital library, as well as functions of Internet database management were given. This is a topic that is continually changing due to the rapid advances that are being made in Internet technology and data management. Next we discussed data mining issues for the web. First we provided some of the challenges in mining Internet databases which include building warehouses as well as mining multimedia databases. Then we discussed how mining could facilitate the user in browsing the web.

Web mining is still a relatively new area and there is active research on this topic. Various conferences are now having panels on web mining (see, for example, [ICTA97]). As web technology and data mining technology mature, we can expect good tools to be developed to mine the large quantities of data on the web. As mentioned earlier, at present many of the data mining tools work on relational databases. However, much of the data on the web is semistructured and unstructured. Therefore, we need to focus our attention on mining text and other types of nonrelational databases. Unless advances are made in this area, successful web mining will be difficult to achieve.

As mentioned earlier, for mining to be effective we need good data. Therefore, to get meaningful results from web mining, we need to have good data on the web. In other words, effective web data management is critical for web mining. There is a lot to be done on web data management. It is only recently that various approaches are being proposed for web data management (see, for example, [IEEE98]). As web data management and data mining technologies mature, we can expect to see good web mining tools to emerge.

CHAPTER 13

SECURITY AND PRIVACY ISSUES FOR DATA MINING

13.1 OVERVIEW

All of our discussions until now have focused on the positive role of data mining. Data mining can be used to improve efficiency, quality of data, marketing and sales, and has many more benefits. Furthermore, even in the case of security problems, we have addressed the case where data mining tools could be used to detect abnormal behavior and intrusions in the system. Data mining also has many applications in detecting fraudulent behavior. While all of these applications of data mining can benefit humans, there is also a dangerous side to mining, since it could be a serious threat to the security and privacy of individuals. This is the topic addressed in this chapter.

One of the challenges to securing databases is the inference problem. Inference is the process of users posing queries and deducing unauthorized information from the legitimate responses that they receive. This problem has been discussed quite a lot over the past two decades. However, data mining makes this problem worse. Users now have sophisticated tools that they can use to get data and deduce patterns that could be sensitive. Without these data mining tools, users would have to be fairly sophisticated in their reasoning to be able to deduce information from posing queries to the databases. That is, data mining tools make the inference problem quite dangerous.

Data mining approaches such as web mining also seriously compromise the privacy of the individuals. One can have all kinds of information about various individuals in a short space of time through browsing the web. Security for digital libraries, Internet databases, and electronic commerce is a subject of much research. Data mining and web mining make this problem even more dangerous. Therefore, protecting the privacy of the individuals is also a major consideration.

This chapter discusses both the inference problem through data mining as well as privacy issues. In Section 13.2, we first provide an overview of the inference problem to give the reader some background. In Section 13.3, we discuss approaches to handling the inference problem that arises through data mining. Since we have discussed quite a bit on warehousing, we also describe data warehousing, inference, and security in this section. Since inductive logic programming is of interest to us, we discuss inference control through the use of inductive logic programming. In Section 13.5, we discuss privacy issues. These privacy issues also depend on policies and procedures enforced. That is, technical, political, as well as social issues play a role here. Then in Section 13.6 we summarize this chapter.

13.2 BACKGROUND ON THE INFERENCE PROBLEM

Inference is the process of posing queries and deducing unauthorized information from the legitimate responses received. For example, the names and salaries of individuals may be unclassified while taken together they are classified. Therefore, one could retrieve names and employee numbers, and then later retrieve the salaries and employee numbers, and make the associations between names and salaries. The problem that occurs through this inference is called the inference problem.

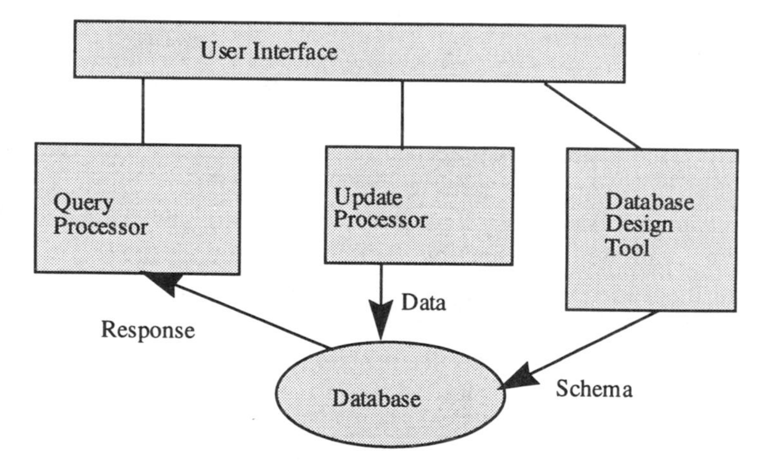

Figure 13-1. Addressing Inference during Query, Update, and Database Design

In the early 1970s, much of the work on the inference problem was on statistical databases. Organizations such as the census bureau were interested in this problem. However, in the mid 1970s and then in the 1980s, the United States Department of Defense started an active research program on multilevel secure databases, and research on the inference problem (see, for example, [AFSB83]) was conducted as part of this effort. The pioneers included Morgenstern [MORG87], Thuraisingham [THUR87], and Hinke [HINK88].

We have conducted extensive research on this subject and worked on various aspects. In particular, it was shown that the general inference problem was unsolvable by Thuraisingham [THUR90b], and then approaches were developed to handle various types of inference. These approaches included those based on security constraints as well as those based on conceptual structures. These approaches handled the inference problem during database design, query, and update operations (see the scenario in Figure 13-1). Furthermore, logic-based approaches were also developed to handle the inference problem (see, for example, [THUR91], [THUR93], and [THUR95]).

Much of the earlier research on the inference problem did not take data mining into consideration. With data mining, users now have tools to make deductions and patterns which could be sensitive. Therefore, in the next section we address inference problem and data mining. We also add some information on data warehousing and inference.

13.3 MINING, WAREHOUSING, AND INFERENCE

First let us give a motivating example where data mining tools are applied to cause security problems. Consider a user who has the ability to apply data mining tools. This user can pose various queries and infer sensitive hypotheses. That is, the inference problem occurs via data mining. This is illustrated in Figure 13-2. There are various ways to handle this problem. Given a database and a particular data mining tool, one can apply the tool to see if sensitive information can be deduced from the unclassified information legitimately obtained. If so, then there is an inference problem. There are some issues with this approach. One is that we are applying only one tool. In reality, the user may have several tools available to him. Furthermore, it is impossible to cover all ways that the inference problem could occur. Some of the security implications are discussed in [CLIF96b].

Another solution to the inference problem is to build an inference controller that can detect the motives of the user and prevent the inference problem from occurring. Such an inference controller lies between the data mining tool and the data source or database, possibly managed by a DBMS. This is illustrated in Figure 13-3. Discussions of security issues for data warehousing and mining are also given in [THUR96b] and [NISS96].

Clifton [CLIF98b] has also conducted some theoretical studies on handling the inference problems that arise through data mining. Clifton's approach is the following. If it is possible to cause doubts in the mind of the adversary that his data mining tool is not a good one, then he will not have confidence in the results. For example, if the classifier built is not a good one for data mining through classification, then the rules produced cannot have sufficient confidence. Therefore, the data mining results also will not have sufficient confidence. Now what are the challenges in making this happen? That is, how can we ensure that the adversary will not have enough confidence in the results? One of the ways is to give only samples of the data to the adversary so that one cannot build a good classifier from these samples (Figure 13-4 illustrates this scenario). The question then is what should the sample be? Clifton has used classification theory to determine the limits of what can be given. This work is still preliminary. There have been some concerns also about this approach, as one could give multiple samples to different groups, and the groups can work together in building a good classifier. But the answer to this is that one needs to keep track of what information is to be given out. At the keynote address on data mining and security (see, for example, [THUR98]), it was suggested that the only way to handle the inference problem is not to give out any samples. But this could mean denial of service. That is, data could be withheld when it is definitely safe to do so.

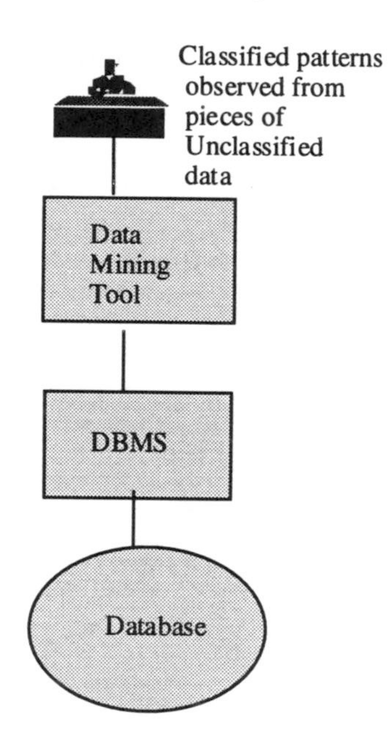

Figure 13-2. Inference Problem

Next, let us focus on data warehousing and inference. We have addressed some security issues for warehouses in [THUR96a]. First of all, security policies of the different data sources that form the warehouse have to be integrated to form a policy for the warehouse. This is not a straightforward task as one has to maintain security rules during the transformations. For example, one cannot give access to an entity in the warehouse, while the same person cannot have access to that entity in the data source. Next, the warehouse security policy has to be enforced. In addition, the warehouse has to be audited. Finally, the inference problem also becomes an issue here. For example, the warehouse may store average salaries. A user can access average salaries and then deduce the individual salaries in the data sources which may be sensitive (see the scenario in Figure 13-5), and therefore, the inference problem could become an issue for the warehouse. To date, little work has been reported on security for data warehouse as well as the inference problem for the warehouse. This is an area that needs much research.

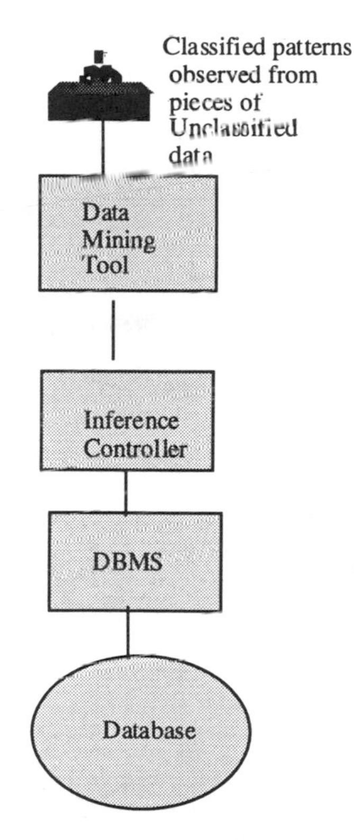

Figure 13-3. Inference Controller

13.4 INDUCTIVE LOGIC PROGRAMMING AND INFERENCE

In the previous section we discussed data mining and the inference problem. In our research we have used deductive logic programming extensively to handle the inference problem. We have specified what we have called security constraints (see, for example, [THUR93]) and then augmented the database system with an inference engine which makes deductions and determines if the constraints are violated. That is, the inference engine, by using the constraints, determines if the new information deduced causes security problems. If this is the case, the data is not released.

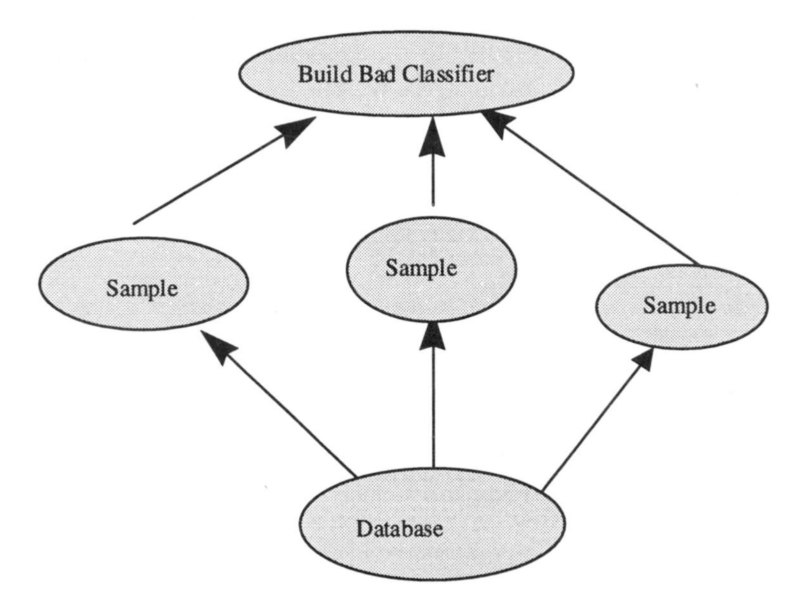

Figure 13-4. Approach to Mining and Inference

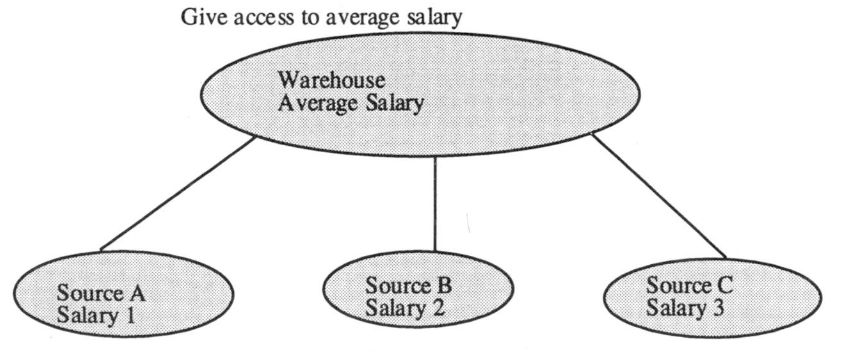

Figure 13-5. Warehousing and Inference

The question is, can this approach be used to control inferences using inductive logic programming techniques? Figure 13-6 is a possible architecture for such an inference controller. This inference controller is based on inductive logic programming. It queries the database, gets the responses, and induces the rules. Some of these rules may be sensitive or lead to giving out sensitive information. Whether the rule is sensitive or can lead to security problems is specified in the form of constraints. If a rule is sensitive, then the inference controller will inform the security officer that the data has potential security problems and may have to be reclassified.

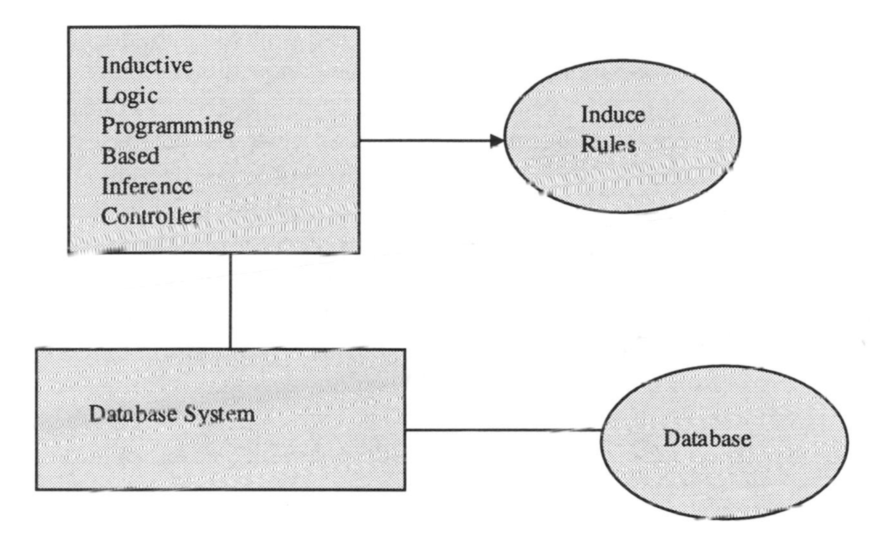

Figure 13-6. Inference Controller based on ILP

In other words, the inference controller that we have mentioned here does not operate on run time. As we have mentioned, it is very difficult to handle all types of data mining tools and prevent users from getting unauthorized information to queries. What the inference controller does is give advice to the security officer regarding potential problems with the data and safety of the data. Some of the issues on inductive inference, which is essentially the technique used in ILP, to handle the inference problem in secure databases is given in [THUR90c].

13.5 PRIVACY ISSUES

At the IFIP (International Federation for Information Processing) working conference on database security in 1997, the group began discussions on privacy issues and the role of web, data mining, and data warehousing (see, for example, [IFIP97]). This discussion continued at the IFIP meeting in 1998 and it was felt that the IFIP group should monitor the developments made by the security working group of the world wide web consortium. The discussions included those based on technical, social, and political aspects (see Figure 13-7). In this section we will examine all of these aspects.

First of all, with the world wide web, there is now an abundance of information about individuals that one can obtain within seconds. This information could be obtained through mining or just from information retrieval. Therefore, one needs to enforce controls on databases and data mining tools. This is a very difficult problem especially with respect to data mining, as we have seen in the previous section. In summary, one needs to develop techniques to prevent users from mining and extracting information from the data whether they be on the web or on servers. Now this goes against all that we have said about mining in the previous chapters. That is, we have portrayed mining as something that is critical for users to have so they can get the right information at the right time. Furthermore, they can also extract patterns previously unknown. This is all true. However, we do not want the information to be used in an incorrect manner. For example, based on information about a person, an insurance company could deny insurance or a loan agency could deny loans. In many cases these denials may not be legitimate. Therefore, information providers have to be very careful in what they release. Also, data mining researchers have to ensure that security aspects are addressed.

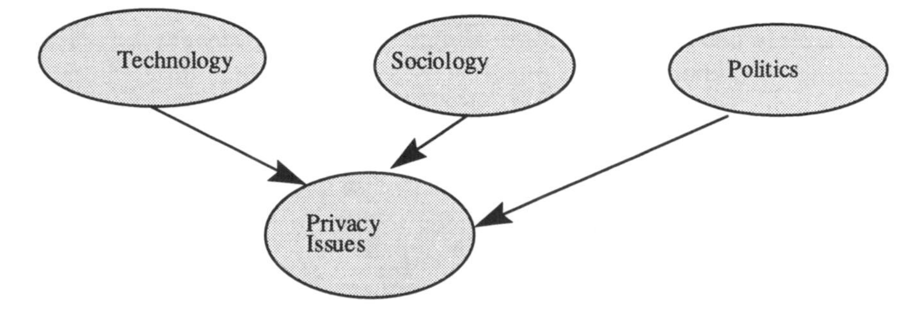

Figure 13-7. Privacy Issues

Next, let us examine the social aspects. In most cultures, privacy of the individuals are important. However, there are certain cultures where it is impossible to ensure privacy. These could be related to political or technological issues or the fact that people have been brought up believing that privacy is not critical. There are places where people divulge their salaries without thinking twice about it, but in many countries, salaries are very private and sensitive. It is not easy to change cultures overnight, and in many cases you do not want to change them as preserving cultures is very important. So what overall effect does this have on data mining and privacy issues? We do not have an answer to this yet as we are only beginning to look into it.

Next, let us examine the political and legal aspects. We include policies and procedures under this. What sort of security controls should one enforce for the web? Should these security polices be mandated or should they be discretionary? What are the consequences of violating the security polices? Who should be administering these policies and managing and implementing them? How is data mining on the web impacted? Can one control how data is mined on the web? Once we have made technological advances on security and data mining, can we enforce security controls on data mining tools? How is information transferred between countries? Again we have no answers to these questions. We have, however, begun discussions. Note that some of the issues we have discussed are related to privacy and data mining, and some others are related to just privacy in general.

We have raised some interesting questions on privacy issues and data mining as well as privacy in general. As mentioned earlier, data mining is a threat to privacy. The challenge is on protecting the privacy but at the same time not losing all the great benefits of data mining. At the 1998 knowledge discovery in databases conference, there was an interesting panel on the privacy issues for web mining. It appears that the data mining as well as the security communities are interested about security and privacy issues. Much of the focus at that panel was on legal issues [KDP98].

13.6 SUMMARY

This chapter is devoted to the important area of security and privacy related to mining. While in Chapter 2 we focused on how data mining could help with security problems, such as auditing and intrusion detection, in this chapter we focused on the negative effects of data mining. In particular, we discussed the inference problem that can result due to mining. First, we gave an overview of the inference problem and then discussed approaches to handle this problem that result from mining. Warehousing and inference issues were also discussed. Then we provided an overview of the privacy issues.

While little work has been reported on security and privacy issues for warehousing and mining, we are moving in the right direction. There is increasing awareness of the problems, and groups such as the IFIP working group in database security are making this a priority. As research initiatives are started in this area, we can expect some progress to be made. Note that there are also social and political aspects to consider. However, we need the technology first before we can enforce various policies and procedures.

CHAPTER 14

METADATA ASPECTS OF MINING

14.1 OVERVIEW

Metadata is a term that has been defined rather loosely. It originates from data management. Initially it was called the data dictionary or the catalog that described the data in the databases [DATE90]. As discussed in Chapter 2, in a relational database the data dictionary will have information about the relations and their attributes. Soon data dictionaries included other information such as access control rules, integrity constraints, and information about data distribution. Subsequently, this information came to be known as metadata. With data warehousing, multimedia, and web information management, the definition of metadata started expanding. Finally, with integration of heterogeneous databases, applications, and systems, metadata is now used synonymously with repository technology.

While we could have included metadata as part of the discussion on technologies addressed in Part I of this book, we chose not to do so since metadata is still a new topic for data mining, and as the technologies are emerging, the definition of metadata is also expanding. Therefore, we have devoted a chapter on metadata under emerging trends. This is an area that will change with technological developments.

The role of metadata in mining is now a subject of much research (see, for example, [META96]). There are two aspects here. One is mining the metadata itself to extract patterns, and the other is to use the metadata to guide the mining process. This chapter will provide a preliminary discussion of metadata mining. First, in Section 14.2, we provide some background information on what metadata is all about for various types of systems. Then, in Section 14.3, we discuss metadata mining. The chapter is summarized in Section 14.4.

14.2 BACKGROUND ON METADATA

Let us revisit the discussion of metadata in Chapter 2. In the example we gave, the database consisted of two relations EMP and DEPT. The metadata includes information about these relations, the number of attributes of each relation, the number of tuples in each relation, and other information such as the creator of the relation (see, for example, Figure 2-17). Now metadata also includes information on the three-schema architecture we discussed in Chapter 2. That is, the external, conceptual, and internal views, as well as the mapping between the three layers, are part of the metadata (see Figure 2-10). In addition, metadata includes information such as "John has read access to EMP and write access to DEPT." Metadata also has information on access methods and index strategies.

Next, let us take it one step further and consider the distributed and heterogeneous databases discussed in Chapter 10. Metadata has information on how the data is distributed. For example, EMP relation may have multiple fragments distributed across multiple sites. Metadata also includes information to handle heterogeneity. For example, the fact that at site 1 an object is interpreted as a ship, and at site 2 it is interpreted as a submarine is part of the metadata. The three-schema architecture discussed in Chapter 2 has been extended to multiple layers to handle heterogeneous schema (see the discussion in [THUR97]). Metadata guides the schema transformation and integration process in handling heterogeneity. Metadata is also needed to migrate legacy databases. Information about the legacy databases is stored as part of the metadata. This metadata is used to transform the legacy database systems to new systems.

We discussed multimedia data management in Chapter 11. Metadata that describes multimedia data could be in different formats. As illustrated in Figure 14-1, in the case of image databases, metadata that describes images could be in text format. Metadata about video and audio data could be in relations or in text. Metadata itself could be multimedia data such as video and audio. Metadata for the web includes information about the various data sources, the locations, the resources on the web, the usage patterns, and the policies and procedures.

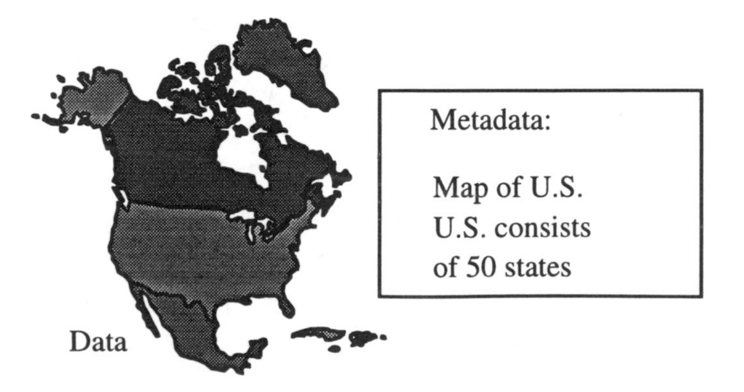

Figure 14-1. Metadata for Image Data

Metadata for data warehousing includes data for integrating the heterogeneous data sources as well as metadata to maintain the warehouse. Metadata can also be generated in the mining process. As data mining is performed, metadata could be gathered about the steps involved in mining. Metadata is also collected for visualization, decision support, machine learning, and statistical reasoning.

In summary, whether it be data mining, data management, web information management, data warehousing, visualization, or decision support, metadata is the central component that is common to all technologies. This is illustrated in Figure 14-2.

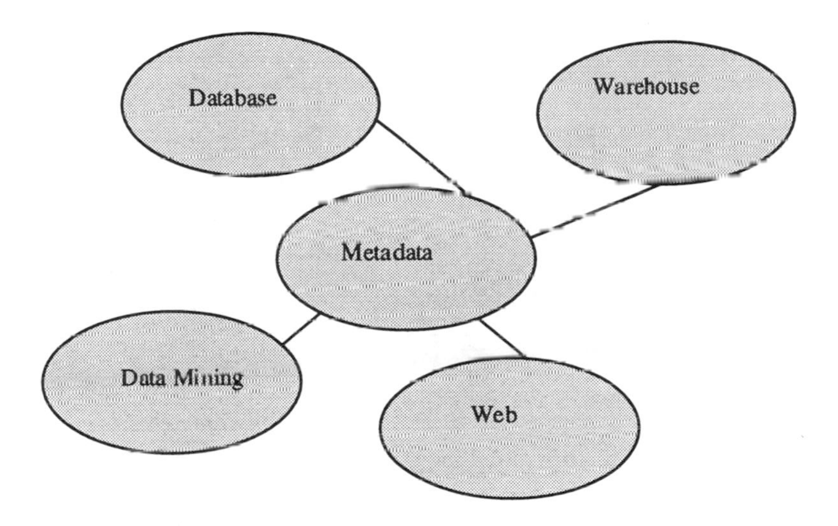

Figure 14-2. Metadata as a Central Component

14.3 MINING AND METADATA

As discussed previously, metadata by itself is becoming a key technology for various tasks such as data management, data warehousing, web searching, multimedia information processing, and now data mining. Because metadata has been so closely aligned with databases in the past, we have included a discussion of the impact of metadata technology on data mining in this book.

Metadata plays an important role in data mining. Metadata could guide the data mining process. That is, the data mining tool could consult the metadatabase and determine the types of queries to pose to the DBMS. Metadata may be updated during the mining process. For example, historical information as well as statistics may be collected during the mining process, and the metadata has to reflect the changes in the environment. The role of metadata in guiding the data mining process is illustrated in Figure 14-3.

There is also another aspect to the role of metadata and that is to conduct data mining on the metadata. Sometimes the data in the database may be incomplete and inaccurate, and the metadata could have more meaningful information. In such a situation, it may be more feasible to mine the metadata and uncover patterns. Mining metadata is illustrated in Figure 14-4.

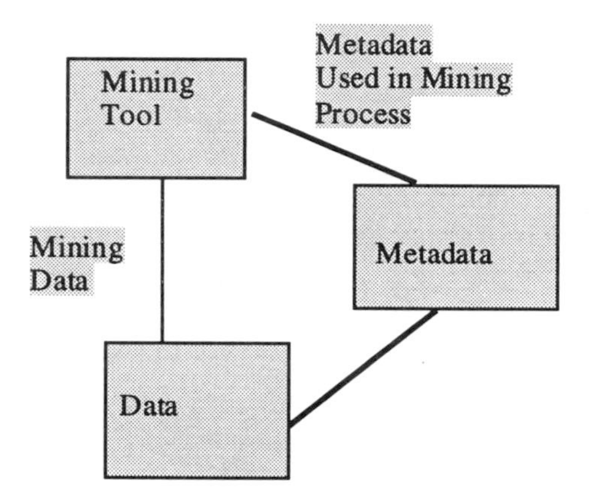

Figure 14-3. Metadata Used in Data Mining

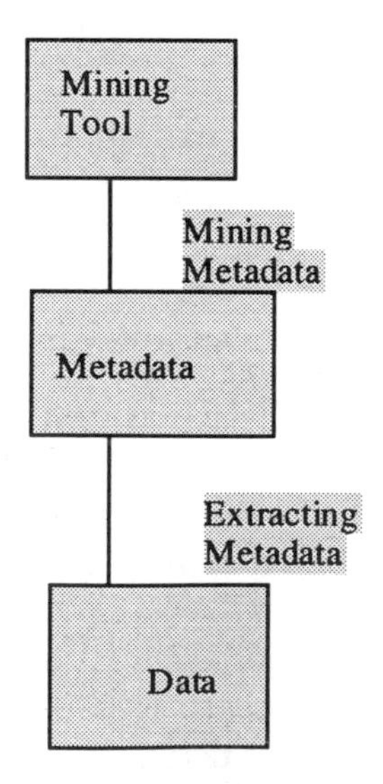

Figure 14-4. Metadata Mining

There has been much discussion recently on the role of metadata for data mining [META96]. There are many challenges here. For example, when is it better to mine the metadata? What are the techniques for metadata mining? How does one structure the metadata to facilitate data mining? Researchers are working on addressing these questions.

Closely associated with the metadata notion is that of a repository. Repository is a database that stores possibly all the metadata, the mappings between various data sources when integrating heterogeneous data sources, information needed to handle semantic heterogeneity such as "ShipX and SubmarineY are the same entity," policies and procedures enforced, as well as information on data quality. So the data mining tool may consult the repository to carry out the mining. On the other hand, the repository itself may be mined. Both these scenarios are illustrated in Figure 14-5.

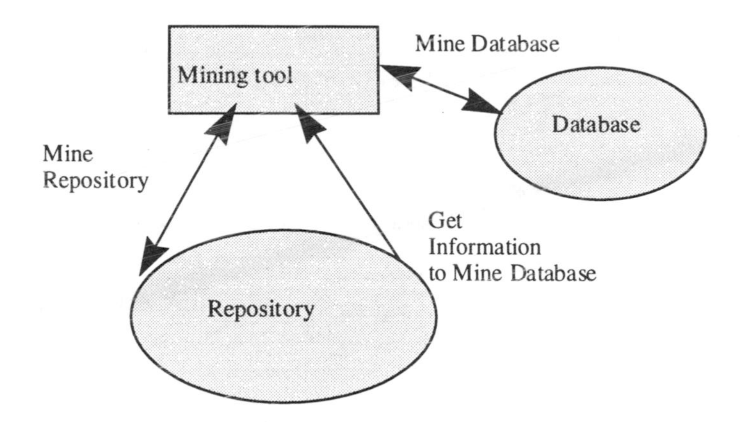

Figure 14-5. Repository and Mining

Metadata plays an important role in various types of mining. For example, in the case of mining multimedia data discussed in Chapter 11, metadata may be extracted from the multimedia databases and then used to mine the data. For example, as illustrated in Figure 14-6, the metadata may help in extracting the key entities from the text. These entities may be mined using commercial data mining tools. Note that in the case of textual data, metadata may include information such as the type of document, the number of paragraphs, and other information describing the document but not the contents of the document itself.

Metadata is also critical in the case of web mining discussed in Chapter 12. Since there is so much information and data on the web, mining this data directly could become quite challenging. Therefore, we may need to extract metadata from the data, and then either mine this metadata or use this metadata to guide in the mining process. This is illustrated in Figure 14-7. Note that languages such as XML that we briefly discussed in Chapter 12 will play a role in describing metadata for web documents.

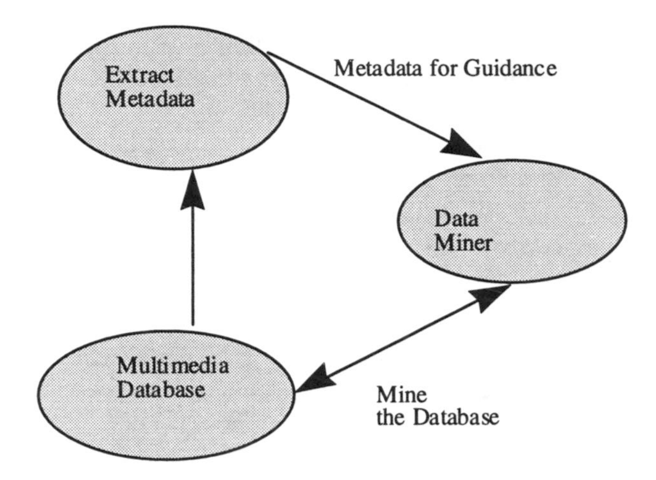

Figure 14-6. Metadata for Multimedia Mining

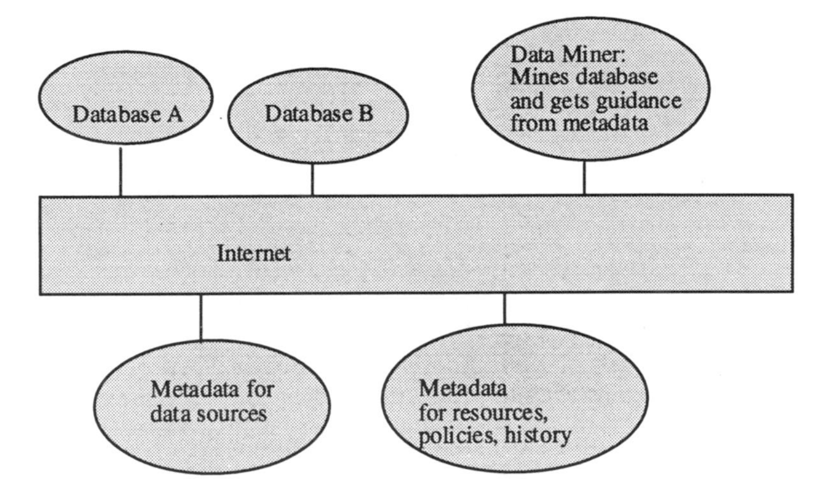

Figure 14-7. Metadata for Web Mining

In Chapter 13 we addressed security and privacy issues for data mining. We mentioned that polices and procedures will be a key issue for determining the extent to which we want to protect the privacy of individuals. These policies and procedures can be regarded as part of the metadata. Therefore, such metadata will have to guide the process of data mining so that privacy issues are not compromised through mining.

In almost every aspect of mining, metadata plays a crucial role. Even in the case of data warehousing, which we have regarded to be a preliminary step to mining, it is important to collect metadata at various stages. For example, in the case of a data warehouse, data from multiple sources have to be integrated. As we discussed in Chapter 3, metadata will guide the transformation process from layer to layer in building the warehouse (see Figure 3-5). Metadata will also help in administering the data warehouse. Also, metadata is used in extracting answers to the various queries posed.

Since metadata is key to all kinds of databases including relational databases, object databases, multimedia databases, distributed, heterogeneous, and legacy databases, and web databases, one could envisage building a metadata repository that contains metadata from the different kinds of databases and then mining the metadata to extract patterns. This approach is illustrated in Figure 14-8 and could be an alternative if the data in the databases are difficult to mine directly.

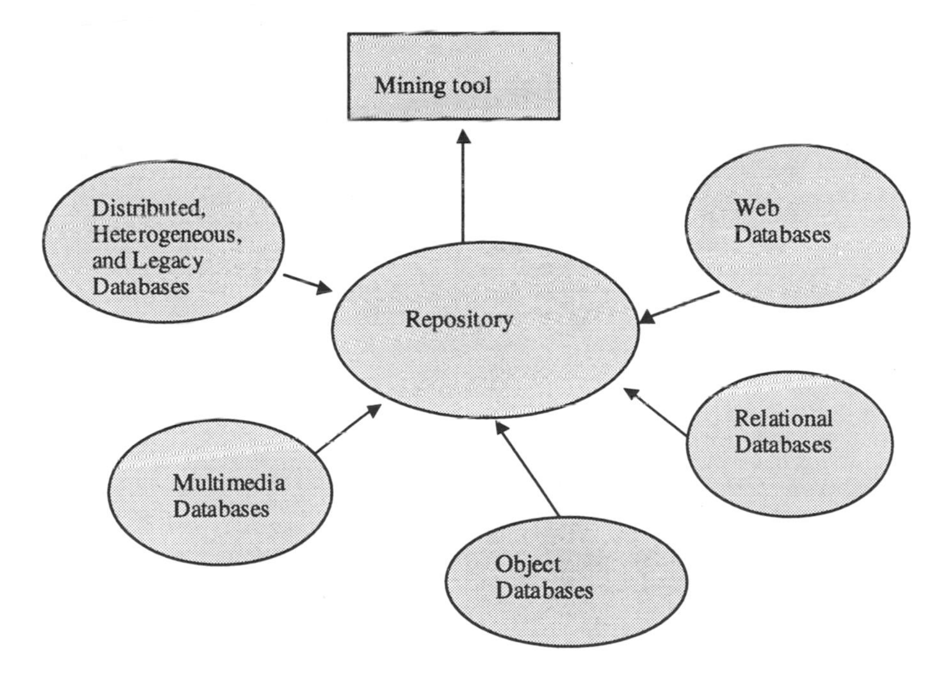

Figure 14-8. Metadata as the Central Repository for Mining

14.4 SUMMARY

This chapter is devoted to a discussion of metadata mining. We first provided an overview of the various types of metadata, and then discussed mining the metadata as well as using metadata to mine the data. Metadata is the central component to many kinds of information systems such as decision support systems, database systems, and machine learning systems.

While metadata can be regarded to be a supporting technology for data mining, we have included metadata mining as part of the discussion on trends because the notion of metadata is continually changing. Initially metadata was considered to be just the data dictionary. Then it included policies, access control rules, and information about data distribution. Now metadata includes information about the various resources on the web, usage patterns, as well as repositories. We believe that metadata mining will become an essential part of all aspects of data mining.

Conclusion to Part III

Part III has discussed some of the emerging trends in mining. These included mining distributed, heterogeneous, and legacy databases, mining multimedia data, web mining, security and privacy issues for mining, and mining metadata. The information in this part, which is Layer 3 of the framework, builds on the information provided by the technologies and concepts in Layers 2 and 3. That is, the chapters in Part III make use of what we have discussed in Parts I and II. We chose these topics as they are considered to be important by many, and they are of interest to us.

The chapters in this book have provided an overview of the technologies, techniques, tools and trends for data mining. This gives the reader some information on what data mining is all about, the technologies involved in supporting mining, and what the directions are. In the concluding chapter, which is Chapter 15, we will discuss the challenges, summarize the directions, and give our view as to where we go from here.

CHAPTER 15

SUMMARY AND DIRECTIONS

15.1 ABOUT THIS CHAPTER

This chapter brings us to closure on this book *Data Mining: Technologies, Techniques, Tools, and Trends*. We have discussed various technologies for data mining, described the concepts and techniques in data mining, and provided an overview of some of the trends in data mining. In particular, we provided an overview of a framework for data mining and then introduced various concepts with respect to this framework. As stated throughout this book, data mining is an integration of multiple technologies, and having good data is key to data mining. This chapter summarizes the contents of this book and then provides an overview of the challenges and directions in data mining. We also give the reader some suggestions on where to go from here.

The organization of this chapter is as follows. In Section 15.2 we summarize the contents of this book. Note that each of the Chapters 2 through 14 gave a summary at the end of the chapter. Essentially we have collected these summaries and put them together to form an overall summary of this book. Then in Section 15.3 we discuss the challenges in data mining. These challenges describe the difficulties that one encounters in doing data mining. Then in Section 15.4 we discuss directions for data mining. Essentially, we consider the trends discussed in Part III of this book, and for each of the topics addressed, we have discussed future work in the area. One could argue that the directions are also part of the challenges. We chose to separate them. In addressing the directions, one needs to address the challenges also. Finally, in Section 15.5 we give suggestions to the reader as to where to go from here. Some of the key points in this book are reiterated, and then we encourage the reader the steps to take to make data mining a success.

15.2 SUMMARY OF THIS BOOK

Chapter 1 provided an introduction to data mining. We first discussed various technologies for data mining. Then we provided an overview of the concepts in data mining. These concepts included the outcomes of mining, the techniques employed, and the approaches used. The directions and trends, such as mining heterogeneous data sources, mining multimedia data, mining web data, metadata aspects, and privacy issues were addressed next. Finally, we illustrated a framework for data mining and showed how we address the components of this framework in this book.

The framework is a three-layer framework. Each layer was described in a part of this book. Part I, which describes the technologies layer, consists of Chapters 2 through 5. Part II, which describes the techniques and tools layer, consists of Chapters 6 through 9, and Part III, which describes the trends layer, consists of Chapters 10 through 14. In the remainder of this section we summarize each of Chapters 2 through 14.

Chapter 2 discussed various aspects of database systems and provided a lot of background information to understand the other chapters in this book. We started with a discussion of various data models. We chose relational, entity-relationship, object-oriented, object relational, and logic-based models, as they are most relevant to what we have addressed in this book. Then we provided an overview of various types of architectures for database systems. These include functional and schema architectures. Next we discussed database design aspects and database administration issues. This chapter also provided a fairly detailed overview of the various functions of database systems. These include query processing, transaction management, storage management, metadata management, security, integrity, and fault tolerance.

After providing an overview of database systems concepts, we then described the issues on integrating data mining with database systems. We discussed various architectures, including both loose and tight coupling approaches. Then we discussed the impact of data modeling on mining such as mining relational databases and object-oriented databases. Next we discussed the impact of data mining on database design and data administration issues. Finally, the impact of data mining on the various database system functions such as query processing, transaction management, metadata management, security and integrity were discussed.

Chapter 3 provided an overview of data warehousing and its relationship to mining. We started with a definition of a data warehouse, the technologies for a data warehouse, functions of a data warehouse, and issues on developing a data warehouse. Key concepts in warehousing include developing a data model, architecture, access methods and index strategies. We then discussed the relationship between warehousing and mining. As mentioned earlier, a frequent question we are asked is, where does warehousing end and mining begin? We discussed our answer to this question. We also discussed the need for warehousing to do mining. Finally, we discussed the relationships between warehousing, database management, and mining.

While Chapters 2 and 3 discussed data management and data warehousing, two key supporting technologies for data mining, Chapter 4 provided an overview of some of the other supporting technologies. These include statistics, machine learning, visualization, parallel processing, and decision support. For each of these supporting technologies we discussed its relationship to mining.

Chapter 5 discussed various architectural aspects for data mining. We started with a discussion of architectures for technology integration. Then we discussed functional architectures for data mining. Finally, we provided an overview of client-server technology for data mining. With this we ended Part I of this book

We started Part II with Chapter 6. This chapter discussed the data mining process from start to finish. We started with a discussion of the various examples so that the reader can have a feel for what data mining can do. These examples were selected from different domains to illustrate different outcomes. These included deviation analyses, correlations and associations, and classifications. Then we discussed the reasons for data mining. Next we discussed the steps to data mining. These included getting the data ready, carrying out mining, pruning the results, identifying actionable items, carrying out the actions, evaluating outcomes, and determining the next cycle. Finally, we discussed the challenges for data mining as well as some user interface aspects.

Chapter 7 explained the concepts in data mining. We started with a discussion of the data mining outcomes, which are also referred to as tasks. These tasks determine what we can expect from data mining. Do we want clustering, classification, or affinity grouping? Then we discussed approaches or methodologies for data mining. For example, once you have determined what outcomes you want, how do you go about doing data mining? Do you take the top-down approach where you start with a hypothesis and then do testing, or do you take a bottom-up approach where you do not start with a hypothesis, or do you take a hybrid approach which is a combination of the two? In the case of the bottom-up approach, do you carry out supervised learning where you direct the learning process, or do you do unsupervised learning where you have no idea what you want to learn at the beginning and start looking for something interesting? Finally, we discussed various data mining techniques such as decision trees, neural networks, and inductive logic programming.

Chapter 8 discussed a particular data mining technique of interest to us. This is called inductive logic programming. It started with some background information on deductive logic programming and then discussed the essential points of inductive logic programming. The use of ILP as a data mining technique and its limitations were also discussed.

Chapter 9 provided a brief overview of the various data mining tools including research prototypes as well as commercial tools. These tools describe the various techniques utilized and the outcomes produced. In describing the research tools, we chose five categories: new functional models, new information services, understandability of the results, scalability, and large-scale projects. Then we described projects in each category. In describing commercial products, we selected five products and explained the essential points for each of them. This ended Part II of this book.

We started Part III with Chapter 10. In this chapter we provided a brief overview of mining distributed, heterogeneous, and legacy databases. This is a relatively new research area and there is little progress to report. However, as data mining tools get more sophisticated, we expect for them to be used on distributed, heterogeneous, and legacy databases. We have pointed out some of the issues and the work that needs to be done.

In Chapter 11 we started with a definition of multimedia database management systems and then provided an overview of these systems. In particular, different types of architectures, data models, and functions of these systems were discussed. Then we addressed data mining for multimedia data. We focused on four types of media: text, image, video, and audio data. We defined what data mining meant for such data and discussed the developments and the challenges. We also provided directions for mining such data.

Chapter 12 discussed the emerging topic of web data mining. First we provided some background on web data management. We have also called this Internet database management or digital libraries. In particular, the definition of a digital library, operation of a digital library, as well as functions of Internet database management were discussed. This is a topic that is continually changing due to the rapid advances being made in Internet technology and data management. Next we discussed data mining issues for the web. First we provided some of the challenges in mining Internet databases, which includes building warehouses as well as mining multimedia databases. Then we discussed how mining could facilitate the user in browsing the web.

Chapter 13 was devoted to the important area of security and privacy related to mining. While in Chapter 2 we focused on how data mining could help with security problems such as auditing and intrusion detection, in this chapter we focused on the negative effects of data mining. In particular, we discussed the inference problem that can result due to mining. First we gave an overview of the inference problem and then discussed approaches to handle this problem. Warehousing and inference issues were also discussed. Then we discussed inductive logic programming and inference. Finally, we provided an overview of the privacy issues.

Chapter 14 was devoted to a discussion of metadata mining. We first provided an overview of the various types of metadata and then discussed mining the metadata as well as using metadata to mine the data. Metadata is the central component to many kinds of information systems such as decision support systems, database systems, and machine learning systems. With this we ended Part III of this book.

15.3 CHALLENGES IN DATA MINING

As stated in Chapter 6, the challenges include having incomplete and inaccurate data, insufficient tools and resources, no management commitment, and continually changing data. The challenges also include mining distributed, heterogeneous, and legacy databases, mining multimedia data, mining web databases, security and privacy of data mining, and metadata aspects of mining. We address each of these challenges.

Much of the data that we collect may be inaccurate, incomplete or uncertain. That is why we have stressed data management and warehousing in this book. In many cases, the data may not be computerized and people may not know where to find the data. So the biggest challenges are to identify the data, computerize it if needed, store it in databases, build warehouses if necessary, clean the data, resolve uncertainty, and format the data so that it can be mined. There is some work going on in various aspects, but we still do not have a good approach to determine how to go about getting good data.

Once we have the data, how do we know what tools to use? The data mining tools are still immature. Often these tools only use one type of data mining technique. The ability to handle multiple techniques for multiple outcomes is a challenge. This is also called multi-stratgey learning in the machine learning community. Therefore, getting good tools is critical as well as knowing when to use the tools.

Other challenges include the ability to prune the results and be able to understand the results. We do not want the user and decision maker to be flooded with data and information. So how do we effectively prune the data mining results? It has been called “needles in the haystack" or "search for the gold nuggets." Some of the research projects are focusing on understanbility of the data. This also includes using various visualization tools for this purpose. Other challenges include multilingual mining which requires tools to handle multiple languages. Scalability of the mining algorithms is also a challenge.

Management support and commitment is also needed to carry out mining. Although this is not a technological challenge, it is an important aspect of mining. Furthermore, before you start a mining project you need to determine whether to contract it out or whether to do it yourself. Since there are not many real world experiences reported on mining, this is a difficult task.

In Part III we discussed various trends in data mining. Each of these trends are also challenges. For example, in the case of distributed and heterogeneous data mining, the question is, do you mine the various databases and then integrate the results, or do you integrate the databases and then do the mining? For legacy databases, the challenge is determining whether it is useful to build tools to mine them since they may not have high quality data. For multimedia databases, one needs to develop tools to mine the multimedia data directly. Currently, many of the mining tools work on relational databases. So do you convert the data in the multimedia databases to relational databases and then mine the data? In doing the conversion you may loose some important information. The area of web mining is just beginning. The challenge is how do you effectively mine the web databases so that useful information is given to the user? The web usage patterns could also be mined to give guidelines to the user in browsing the web. Security and privacy are critical areas. While data mining is such a useful tool for decision makers, if used improperly it can invade the privacy of individuals. The challenge here is how to protect the privacy of the individuals from the numerous mining tools out there. Finally, metadata is becoming an increasingly important resource for various aspects of data management. The challenge here is how to effectively use metadata for mining.

This book has discussed many challenges. We have answers to a few. There is still much to be done on data mining so that decision makers can benefit from it. The next section discusses some of the directions.

15.4 DIRECTIONS FOR DATA MINING

Now that we have described the challenges, let us provide some directions. First, we need various research agendas to make data mining a success. We need inputs from multiple communities; however, these communities have to work together. Therefore, we need collaborative research projects between data management, machine learning, statistics, and other communities for this. Even with vendors, we need multiple vendors to integrate their tools to produce good data mining tools. We gave some examples in Chapter 9 where data warehousing and mining tools were integrated. Several areas need further work. Here are some.

- Data Understanding: Here we need to focus on understanding the results of mining and determining what information is useful and what to prune.
- Incomplete and Uncertain Data: Data Warehousing and data management work should be expanded to include techniques to handle data that may not be complete or correct. Data quality plays a role here.
- Multilingual Mining: Data mining now works on text in a single language. For many applications we may need to mine multiple languages. This is especially important for journalists.
- Multi-strategy Learning: While many developments have been made on learning techniques, we need tools to handle multiple learning techniques and strategies. Such tools then need to adapt to the various situations.
- Scalability: As we have stressed, data mining has to work on large databases. Presently, we have even peta-byte sized databases. Can the techniques scale for these databases? We need to conduct theoretical as well as simulation studies and also test with larger and larger data sets.
- Better Data Mining Techniques: In general we need to improve on the data mining techniques. We need to study the limitations of the techniques and determine which techniques can be used for which applications and outcomes.
- Theory of Data Mining: At present, data mining is still an art. Therefore, to better understand mining, we need to develop a theory.
- Integration of Technologies: Integrating multiple technologies effectively is key to good mining. We have discussed some of the issues. We need better tools for integration.

- Distributed, Heterogeneous, and Legacy Database Mining: We need techniques to handle databases that are distributed, heterogeneous, and legacy in nature.
- Multimedia Data Mining: We need to develop tools to mine multimedia data directly. If multimedia data is to be transformed into relational databases, then we need techniques to transform the data so that important information is not lost. We also need tools to mine object databases since some of the multimedia data are stored in object databases.
- Web Mining: Extracting only the useful information from the web for the user and guiding the user to navigate the web are two important areas. Some progress has been reported. We need better tools for this.
- Metadata Mining: We need to identify the metadata that can be mined and that can be used for mining. Do you need special tools for metadata or can you use the tools developed for data mining?
- Security and Privacy of Data Mining: We need technology advances as well as policies and procedures for this. It is almost impossible to protect from all kinds of mining tools. We need to understand the limitations of the approaches.

15.5 WHERE DO WE GO FROM HERE?

Some key points we have stated are the following. Data mining is not a stand-alone technology. It is an integration of multiple technologies. Therefore, technologies such as data management, machine learning, statistical reasoning, high performance computing, decision support, and visualization have to work together to make data mining a success. Another point is that having good data is key to mining. More work has to be done in organizing, cleaning, and structuring the data. Data mining is still in its infancy. There is a lot to do, many challenges to overcome, and new problems to solve. A final point is that data mining is still very much an art at present. As more developments are made we can expect formalisms to be developed. Furthermore, we can also expect various methodologies and object-oriented design and analysis techniques such as the Unified Modeling Language (UML) to be applied for mining. For a discussion of UML we refer to [FOWL97].

We have provided a broad overview of data mining and discussed technologies, techniques, tools and trends. We have also given many references should the reader need in-depth coverage of a particular topic. However, all the reading is not going to give the reader a better appreciation for what data mining is all about. It is certainly useful to have a good knowledge in data mining and be able to speak intelligently about it. However, if you want to know what technique works for what application, what are the limitations of an algorithm, or how do you want to train your tool, then you need hands-on experience with the tools. That is, since data mining in its present form gets better with practice, we urge the reader to work with practical applications in using the data mining tools as well as with developing the tools.

Another point to note is that when you want to start a data mining project, you need management buy-in. This means financial and personnel resources. Furthermore, you need to decide whether to contract the work or have it done in-house. If you are using a commercial tool, then you need to have frequent communication with the developer. In other words, the customer, the data mining tool developer, and those who do the data mining have to work very closely together, or else the project may be a failure.

This book has also given some brief information about products and prototypes. As we have stressed, we have selected these products only because of our familiarity with them. We are not endorsing any of the products. Furthermore, due to the rapid developments in the field, the information about these products may soon be outdated. Therefore, we urge the reader to take advantage of the various commercial and research material available on these products and prototypes.

We believe that there are exciting opportunities in data mining with the emergence of new technologies such as warehousing, web information management, and multimedia data management. Furthermore, technology integration, such as integration of data management and machine learning, is making a lot of progress. As the user gets flooded with more and more data and information, the need to analyze this information, give only the information the user needs, and extract previously unknown information to help the user in decision making process will become urgent. That is, there will be a critical need for data mining, and the demand for data mining experts will continue to grow. We feel that the opportunities and challenges in data management, in general, and data mining in particular, will be endless.

References

REFERENCES

[ACM90] Special Issue on Heterogeneous Database Systems, ACM Computing Surveys, September 1990.

[ACM91] Special Issue on Next Generation Database Systems, Communications of the ACM, October 1991.

[ACM95] Special Issue on Digital Libraries, Communications of the ACM, May 1995.

[ACM96a] Special Issue on Data Mining, Communications of the ACM, November 1996.

[ACM96b] Special Issue on Electronics Commerce, Communications of the ACM, June 1996.

[ADRI96] Adriaans, P., and Zantinge, D., "Data Mining," Addison Wesley, MA, 1996.

[AFCE97] Proceedings of the First Federal Data Mining Symposium, Washington D.C., December 1997.

[AFSB83] Air Force Summer Study Board Report on Multilevel Secure Database Systems, Department of Defense Document, 1983.

[AGRA93] Agrawal, A. et al.., "Database Mining a Performance Perspective," IEEE Transactions on Knowledge and Data Engineering, Vol. 5, December 1993.

[BANE87] Banerjee, J. et al., "A Data Model for Object-Oriented Applications," ACM Transactions on Office Information Systems, Vol. 5, 1987.

[BELL92] Bell D. and Grimson, J., "Distributed Database Systems," Addison Wesley, MA, 1992.

[BENS95] Bensley, E. et al., "Evolvable Systems Initiative for Real-time Command and Control Systems," Proceedings of the 1st IEEE Complex Systems Conference, Orlando, FL, November 1995.

[BERN87] Bernstein, P. et al., "Concurrency Control and Recovery in Database Systems," Addison Wesley, MA, 1987.

[BERR97] Berry, M. and Linoff, G., "Data Mining Techniques for Marketing, Sales, and Customer Support," John Wiley, NY, 1997.

[BROD84] Brodie, M. et al., "On Conceptual Modeling: Perspectives from Artificial Intelligence, Databases, and Programming Languages," Springer Verlag, NY, 1984.

[BROD86] Brodie, M. and Mylopoulos, J., "On Knowledge Base Management Systems," Springer Verlag, NY, 1986.

[BROD88] Brodie, M. et al., "Readings in Artificial Intelligence and Databases," Morgan Kaufmann, CA, 1988.

[BROD95] Brodie M. and Stonebraker, M., "Migrating Legacy Databases," Morgan Kaufmann, CA, 1995.

[BUNE82] Buneman, P., "Functional Data Model," ACM Transactions on Database Systems, 1983.

[CARB98] Carbone, P., "Data Mining," Handbook of Data Management, Auerbach Publications, NY, 1998 (Ed: B. Thuraisingham).

[CERI84] Ceri, S. and Pelagatti, G., "Distributed Databases, Principles and Systems," McGraw Hill, NY, 1984.

[CHAN73] Chang C., and Lee R., "Symbolic Logic and Mechanical Theorem Proving," Academic Press, NY, 1973.

[CHEN76] Chen, P., "The Entity Relationship Model - Toward a Unified View of Data," ACM Transactions on Database Systems, Vol. 1, 1976.

[CHOR94] Chorafas, D., "Intelligent Multimedia Databases," Prentice Hall, NJ, 1994.

[CLIF96a] Clifton, C, and Morey, D., "Data Mining Technology Survey," Private Communication, Bedford, MA, December 1996.

[CLIF96b] Clifton, C. and Marks, D., "Security and Privacy Issues for Data Mining," Proceedings of the ACM SIGMOD Conference Workshop on Data Mining, Montreal, Canada, June 1996.

[CLIF98a] Clifton, C., "Image Mining," Private Communication, Bedford, MA, July 1998.

[CLIF98b] Clifton C., "Privacy Issues for Data Mining," Private Communication, Bedford, MA, April 1998.

[CODD70] Codd, E. F., "A Relational Model of Data for Large Shared Data Banks," Communications of the ACM, Vol. 13, 1970.

[COOL98] Cooley, R., "Taxonomy for Web Mining," Private Communication, Bedford, MA, August 1998.

[DARPA98] Workshop on Knowledge Discovery in Databases, Defense Advanced Research Projects Agency, Pittsburgh, PA, June 1998.

[DAS92] Das, S., "Deductive Databases and Logic Programming," Addison Wesley, MA, 1992.

[DATE90] Date, C. J., "An Introduction to Database Management Systems," Addison Wesley, MA, 1990 (6th edition published in 1995 by Addison Wesley).

[DCI96] Proceedings of the DCI Conference on Databases and Client Server Computing, Boston, MA, March 1996.

[DE98] Proceedings of the 1998 Data Engineering Conference, Orlando, FL, February 1998.

[DECI] Decision Support Journal, Elsevier/North Holland Publications.

[DEGR86] DeGroot, T., "Probability and Statistics," Addison Wesley, MA, 1986.

[DIGI95] Proceedings of the Advances in Digital Libraries Conference, McLean, VA, May 1995, (Ed: N. Adam et al.).

[DIST98] Workshop on Distributed and Parallel Data Mining, Melbourne, Australia, April 1998.

[DMH94] Data Management Handbook, Auerbach Publications, NY, 1994 (Ed: B. von Halle and D. Kull).

[DMH95] Data Management Handbook Supplement, Auerbach Publications, NY, 1995 (Ed: B. von Halle and D. Kull).

[DMH96] Data Management Handbook Supplement, Auerbach Publications, NY, 1996 (Ed: B. Thuraisingham).

[DMH98] Data Management Handbook Supplement, Auerbach Publications, NY, 1998 (Ed: B. Thuraisingham).

[DOD94] Proceedings of the 1994 DoD Database Colloquium, San Diego, CA, August 1994.

[DOD95] Proceedings of the 1994 DoD Database Colloquium, San Diego, CA, August 1995.

[DSV98] DSV Laboratory, "Inductive Logic Programming," Private Communication, Stockholm, Sweden, June 1998.

[FAYY96] Fayyad, U. et al., "Advanced in Knowledge Discovery and Data Mining," MIT Press, MA, 1996.

[FELD95] Feldman, R. and Dagan, I., "Knowledge Discovery in Textual Databases (KDT)," Proceedings of the 1995 Knowledge Discovery in Databases Conference, Montreal, Canada, August 1995.

[FOWL97] Fowler, M. et al., "UML Distilled: Applying the Standard Object Modeling Language," Addison Wesley, MA, 1997.

[FROS86] Frost, R., "On Knowledge Base Management Systems," Collins Publishers, U.K., 1986.

[GALL78] Gallaire, H. and Minker, J., "Logic and Databases," Plenum Press, NY, 1978.

[GRIN95] Grinstein, G. and Thuraisingham, B., "Data Mining and Visualization: A Position Paper," Proceedings of the Workshop on Databases in Visualization, Atlanta GA, October 1995.

[GRUP98] Grupe F. and Owrang, M., "Database Mining Tools", in the Handbook of Data Management Supplement, Auerbach Publications, NY, 1998 (Ed: B.Thuraisingham).

[HAN98] Han, J., "Data Mining," Keynote Address, Second Pacific Asia Conference on Data Mining, Melbourne, Australia, April 1998.

[HINK88] Hinke T., "Inference and Aggregation Detection in Database Management Systems," Proceedings of the 1988 Conference on Security and Privacy, Oakland, CA, April 1988.

[ICTA97] Panel on Web Mining, International Conference on Tools for Artificial Intelligence, Newport Beach, CA, November 1997.

[IEEE89] "Parallel Architectures for Databases," IEEE Tutorial, 1989 (Ed: A. Hurson et al.).

[IEEE91] Special Issue in Multidatabase Systems, IEEE Computer, December 1991.

[IEEE98] IEEE Data Engineering Bulletin, June 1998.

[IFIP] Proceedings of the IFIP Conference Series in Database Security, North Holland.

[IFIP97] "Web Mining," Proceedings of the 1997 IFIP Conference in Database Security, Lake Tahoe, CA, August 1997..

[ILP97] Summer School on Inductive Logic Programming, Prague, Czech Republic, September 1998.

[INMO93] Inmon, W., "Building the Data Warehouse," John Wiley and Sons, NY, 1993.

[JUNG98] Junglee Corporation, "Virtual Database Technology, XML, and the Evolution of the Web," IEEE Data Engineering Bulletin, June 1998 (authors: Prasad and Rajaraman).

[KDD95] Proceedings of the First Knowledge Discovery in Databases Conference, Montreal, Canada, August 1995.

[KDD96] Proceedings of the Second Knowledge Discovery in Databases Conference, Portland, OR, August 1996.

[KDD97] Proceedings of the Third Knowledge Discovery in Databases Conference, Newport Beach, CA, August 1997.

[KDD98] Proceedings of the Fourth Knowledge Discovery in Databases Conference, New York, NY, August 1998.

[KDP98] Panel on Privacy Issues for Data Mining, Knowledge Discovery in Databases Conference, New York, NY, August 1998.

[KDT98] Tutorial on Commercial Data Mining Tools, Knowledge Discovery in Databases Conference, August 1998 (Presenters: J. Elder and D. Abbott)

[KIM85] Kim, W. et al., “Query Processing in Database Systems,” Springer Verlag, NY, 1985.

[KORT86] Korth, H. and Silberschatz, A., "Database System Concepts," McGraw Hill, NY, 1986.

[KOWA74] Kowalski, R. A., "Predicate Logic as a Programming Language," Information Processing 74, Stockholm, North Holland Publications, 1974.

[LIN97] Lin, T.Y., (Editor) “Rough Sets and Data Mining,” Kluwer Publishers, MA, 1997.

[LLOY87] Lloyd, J., “Foundations of Logic Programming,” Springer Verlag, Germany, 1987.

[LOOM95] Loomis, M., “Object Databases,” Addison Wesley, MA, 1995.

[MAIE83] Maier, D., “Theory of Relational Databases,” Computer Science Press, MD, 1983.

[MATTO98] Mattox, D. et al., "Software Agents for Data Management," Handbook of Data Management, Auerbach Publications, NY, 1998 (Ed: B. Thuraisingham).

[MDDS94] Proceedings of the Massive Digital Data Systems Workshop, published by the Community Management Staff, Washington D.C., 1994.

[MERL97] Merlino, A. et al., "Broadcast News Navigation using Story Segments," Proceedings of the 1997 ACM Multimedia Conference, Seattle, WA, November 1998.

[META96] Proceedings of the 1st IEEE Metadata Conference, Silver Spring, MD, April 1996 (Originally published on the web, Editor: R. Musick, Lawrence Livermore National Laboratory).

[MINK88] Minker, J., (Editor) "Foundations of Deductive Databases and Logic Programming," Morgan Kaufmann, CA, 1988 (Editor).

[MIT] Technical Reports on Data Quality, Sloan School, Massachusetts Institute of Technology, Cambridge, MA.

[MITC97] Mitchell, T., "Machine Learning," McGraw Hill, NY, 1997.

[MORE98a] Morey, D., "Knowledge Management Architecture," Handbook of Data Management, Auerbach Publications, NY, 1998 (Ed: B. Thuraisingham).

[MORE98b] Morey, D., "Web Mining," Private Communication, Bedford, MA, June 1998.

[MORG88] Morgenstern, M., "Security and Inference in Multilevel Database and Knowledge Base Systems," Proceedings of the 1987 ACM SIGMOD Conference, San Francisco, CA, June 1987.

[NG97] Ng, R., "Image Mining," Private Communication, Vancouver, British Columbia, December 1997.

[NISS96] Panel on Data Warehousing, Data Mining, and Security, Proceedings of the 1996 National Information Systems Security Conference, Baltimore, MD, October 1996.

[NISS97] Papers on Internet Security, Proceedings of the 1997 National Information Systems Conference, Baltimore, MD, October 1997.

[NSF90] Proceedings of the Database Systems Workshop, Report published by the National Science Foundation, 1990 (also in ACM SIGMOD Record, December 1990).

[NSF95] Proceedings of the Database Systems Workshop, Report published by the National Science Foundation, 1995 (also in ACM SIGMOD Record, March 1996).

[NWOS96] Nwosu, K. et al., (Editors) "Multimedia Database Systems, Design and Implementation Strategies." Kluwer Publications, MA, 1996.

[ODMG93] "Object Database Standard: ODMB 93," Object Database Management Group, Morgan Kaufmann, CA, 1993.

[OMG95] "Common Object Request Broker Architecture and Specification," OMG Publications, John Wiley, NY, 1995.

[ORFA94] Orfali, R. et al., "Essential, Client Server Survival Guide," John Wiley, NY, 1994.

[ORFA96] Orfali, R. et al., "The Essential, Distributed Objects Survival Guide," John Wiley, NY, 1994.

[PAKDD97] Proceedings of the Knowledge Discovery in Databases Conference, Singapore, February 1997.

[PAKDD98] Proceedings of the Second Knowledge Discovery in Databases Conference, Melbourne, Australia, April 1998.

[PRAB97] Prabhakaran, B., "Multimedia Database Systems," Kluwer Publications, MA, 1997.

[QUIN93] Quinlan, R., "C4.5: Programs for Machine Learning," Morgan Kaufmann, CA, 1993.

[RAMA94] Ramakrishnan, R., (Editor) Applications of Deductive Databases, Kluwer Publications, MA, 1994.

[ROSE98] Rosenthal, A., "Multi-Tier Architecture," Private Communication, Bedford, MA, August 1998.

[SIGM96] Proceedings of the ACM SIGMOD Workshop on Data Mining, Montreal, Canada, May 1996.

[SIGM98] Proceedings of the 1998 ACM SIGMOD Conference, Seattle, WA, June 1998.

[SIMO95] Simoudis, E. et al., "Recon Data Mining System," Technical Report, Lockheed Martin Corporation, 1995.

[SQL3] "SQL3," American National Standards Institute, Draft, 1992 (a version also presented by J. Melton at the Department of Navy's DISWG NGCR meeting, Salt Lake City, UT, November 1994).

[STAN98] Stanford Database Group Workshop, Jungalee Virtual Relational Database, September 1998 (also appeared in IEEE Data Engineering Bulletin, June 1998).

[THUR87] Thuraisingham, B., "Security Checking in Relational Database Systems Augmented by an Inference Engine," Computers and Security, Vol. 6, 1987.

[THUR90a] Thuraisingham, B., "Nonmonotonic Typed Multilevel Logic for Multilevel Secure Database Systems," MITRE Report, June 1990 (also published in the Proceedings of the1992 Computer Security Foundations Workshop, Franconia, NH, June 1991).

[THUR90b] Thuraisingham, B., "Recursion Theoretic Properties of the Inference Problem," MITRE Report, June 1990 (also presented at the 1990 Computer Security Foundations Workshop, Franconia, NH, June 1990).

[THUR90c] Thuraisingham, B., "Novel Approaches to Handle the Inference Problem," Proceedings of the 1990 RADC Workshop in Database Security, Castile, NY, June 1990.

[THUR91] Thuraisingham, B., "On the Use of Conceptual Structures to Handle the Inference Problem," Proceedings of the 1991 IFIP Database Security Conference, Shepherdstown, WVA, November 1991.

[THUR93] Thuraisingham, B. et al., "Design and Implementation of a Database Inference Controller," Data and Knowledge Engineering Journal, North Holland, Vol. 8, December 1993.

[THUR95] Thuraisingham, B. and Ford, W., "Security Constraint Processing in a Multilevel Secure Distributed Database Management System," IEEE Transactions on Knowledge and Data Engineering, Vol. 7, 1995.

[THUR96a] Thuraisingham, B., "Data Warehousing, Data Mining, and Security (Version 1)," Proceedings of the 10th IFIP Databasc Sccurity Conference, Como, Italy, 1996.

[THUR96b] Thuraisingham, B., "Internet Database Management," Database Management, Auerbach Publications, NY, 1996.

[THUR96c] Thuraisingham, B., "Interactive Data Mining and the World Wide Web," Proceedings of Compugraphics Conference, Paris, France, December 1996.

[THUR97] Thuraisingham, B., " Data Management Systems Evolution and Interoperation," CRC Press, FL, May 1997.

[THUR98] Thuraisingham, B., "Data Warehousing, Data Mining, and Security (Version 2)," Keynote Address at Second Pacific Asia Conference on Data Mining, Melbourne, Australia, April 1998.

[TKDE93] Special Issue on Data Mining, IEEE Transactions on Knowledge and Data Engineering, December 1993.

[TKDE96] Special Issue on Data Mining, IEEE Transactions on Knowledge and Data Engineering, December 1996.

[TRUE89] Trueblood, R. and Potter, W., "Hyper-Semantic Data Modeling," Data and Knowledge Engineering Journal, Vol. 4, North Holland, 1989.

[TSUR98] Tsur, D. et al., "Query Flocks: A Generalization of Association Rule Mining," Proceedings of the 1998 ACM SIGMOD Conference, Seattle, WA, June 1998.

[TSIC82] Tsichritzis, D. and Lochovsky, F., "Data Models," Prentice Hall, NJ, 1982.

[ULLM88] Ullman, J. D., "Principles of Database and Knowledge Base Management Systems," Volumes I and II, Computer Science Press, MD 1988.

[VIS95] Proceedings of the 1995 Workshop on Visualization and Databases, Atlanta, GA, October 1997 (Ed: G. Grinstein)

[VIS97] Proceedings of the 1997 Workshop on Visualization and Data Mining, Phoenix, AZ, October 1997 (Ed: G. Grinstein).

[VLDB98] Proceedings of the Very Large Database Conference, New York City, NY, August 1998.

[WIED92] Wiederhold, G., "Mediators in the Architecture of Future Information Systems," IEEE Computer, March 1992.

[WOEL86] Woelk, D. et al., "An Object-Oriented Approach to Multimedia Databases," Proceedings of the ACM SIGMOD Conference, Washington DC, June 1986.

[XML98] Extended Markup Language, Document published by the World Wide Web Consortium, Cambridge, MA, February 1998.

Appendices

APPENDIX A

DATA MANAGEMENT TECHNOLOGY

A.1 OVERVIEW

In this appendix we provide an overview of the developments and trends in data management as discussed in our previous book *Data Management Systems Evolution and Interoperation* [THUR97]. Since data plays a major role in data mining, a good understanding of data management is essential for data mining.

Recent developments in information systems technologies have resulted in computerizing many applications in various business areas. Data has become a critical resource in many organizations and therefore efficient access to data, sharing the data, extracting information from the data, and making use of the information have become urgent needs. As a result, there have been several efforts on integrating the various data sources scattered across several sites. These data sources may be databases managed by database management systems or they could simply be files. To provide the interoperability between the multiple data sources and systems, various tools are being developed. These tools enable users of one system to access other systems in an efficient and transparent manner.

We define data management systems to be systems that manage the data, extract meaningful information from the data, and make use of the information extracted. Therefore, data management systems include database systems, data warehouses, and data mining systems. Data could be structured data such as that found in relational databases or it could be unstructured such as text, voice, imagery, and video. There have been numerous discussions in the past to distinguish between data, information, and knowledge.[21] We do not attempt to clarify these terms. For our purposes, data could be just bits and bytes or it could convey some meaningful information to the user. We will, however, distinguish between database systems and database management systems. A database management system is that component which manages the database containing persistent data. A database system consists of both the database and the database management system.

---

[21] More recently the area of knowledge management is receiving a lot of attention. We have not addressed knowledge management in this book. For details we refer to [MORE98].

A key component to the evolution and interoperation of data management systems is the interoperability of heterogeneous database systems. Efforts on the interoperability between database systems were reported since the late 1970s. However, it is only recently that we are seeing commercial developments in heterogeneous database systems. Major database system vendors are now providing interoperability between their products and other systems. Furthermore, many of the database system vendors are migrating towards an architecture called the client-server architecture which facilitates distributed data management capabilities. In addition to efforts on the interoperability between different database systems and client-server environments, work is also directed towards handling autonomous and federated environments.

The organization of this appendix is as follows. Since database systems are a key component of data management systems, we first provide an overview of the developments in database systems. These developments are discussed in Section A.2. Then we provide a vision for data management systems in Section A.3. Our framework for data management systems is discussed in Section A.4. Note that data mining as well as warehousing are components of this framework. Building information systems from our framework with special instantiations is discussed in Section A.5. This appendix is summarized in Section A.6. References are given in Section A.7.

A.2 DEVELOPMENTS IN DATABASE SYSTEMS

Figure A-1 provides an overview of the developments in database systems technology. While the early work in the 1960s focused on developing products based on the network and hierarchical data models, much of the developments in database systems took place after the seminal paper by Codd describing the relational model [CODD70]. Research and development work on relational database systems was carried out during the early 1970s and several prototypes were developed throughout the 1970s. Notable efforts include IBM's (International Business Machine Corporation's) System R and University of California at Berkeley's Ingres. During the 1980s, many relational database system products were being marketed (notable among these products are those of Oracle Corporation, Sybase Inc., Informix Corporation, Ingres Corporation, IBM, Digital Equipment Corporation, and Hewlett Packard Company). During the 1990s, products from other vendors have emerged (e.g., Microsoft Corporation). In fact, to date numerous relational database system products have been marketed. However, Codd has stated that many of the systems that are being marketed as relational systems are not really relational (see, for example, the discussion in [DATE90]). He then discussed various criteria that a system must satisfy to be qualified as a relational database system. While the early work focused on issues such as data model, normalization theory, query processing and optimization strategies, query languages, and access strategies and indexes, later the focus shifted toward supporting a multi-user environment. In particular, concurrency control and recovery techniques were developed. Support for transaction processing was also provided.

Research on relational database systems as well as on transaction management, was followed by research on distributed database systems around the mid-1970s. Several distributed database system prototype development efforts also began around the late 1970s. Notable among these efforts include IBM's System R*, DDTS (Distributed Database Testbed System) by Honeywell Inc., SDD-I and Multibase by CCA (Computer Corporation of America), and Mermaid by SDC (System Development Corporation). Furthermore, many of these systems (e.g., DDTS, Multibase, Mermaid) function in a heterogeneous environment. During the early 1990s several database system vendors (such as Oracle Corporation, Sybase Inc., Informix Corporation) provided data distribution capabilities for their systems. Most of the distributed relational database system products are based on client-server architectures. The idea is to have the client of vendor A communicate with the server database system of vendor B. In other words, the client-server computing paradigm facilitates a heterogeneous computing environment. Interoperability between relational and non-relational commercial database systems is also possible. The database systems community is also involved in standardization efforts. Notable among the standardization efforts are the ANSI/SPARC 3-level schema architecture,[22] the IRDS (Information Resource Dictionary System) standard for Data Dictionary Systems, the relational query language SQL (Structured Query Language), and the RDA (Remote Database Access) protocol for remote database access.

Another significant development in database technology is the advent of object-oriented database management systems. Active work on developing such systems began in the mid-1980s and they are now commercially available (notable among them include the products of Object Design Inc., Ontos Inc., Gemstone Systems Inc., Versant Object Technology). It was felt that new generation applications such as

---

22 ANSI stands for American National Standards Institute. SPARC stands for Systems Planning and Requirements Committee.

multimedia, office information systems, CAD/CAM,[23] process control, and software engineering have different requirements. Such applications utilize complex data structures. Tighter integration between the programming language and the data model is also desired. Object-oriented database systems satisfy most of the requirements of these new generation applications [CATT91].

According to the Lagunita report published as a result of a National Science Foundation (NSF) workshop in 1990 [NSF90], relational database systems, transaction processing, and distributed (relational) database systems are stated as mature technologies. Furthermore, vendors are marketing object-oriented database systems and demonstrating the interoperability between different database systems. The report goes on to state that as applications are getting increasingly complex, more sophisticated database systems are needed. Furthermore, since many organizations now use database systems, in many cases of different types, the database systems need to be integrated. Although work has begun to address these issues and commercial products are available, several issues still need to be resolved. Therefore, challenges faced by the database systems researchers in the early 1990s were in two areas. One is next generation database systems and the other is heterogeneous database systems.

Next generation database systems include object-oriented database systems, functional database systems, special parallel architectures to enhance the performance of database system functions, high performance database systems, real-time database systems, scientific database systems, temporal database systems, database systems that handle incomplete and uncertain information, and intelligent database systems (also sometimes called logic or deductive database systems).[24] Ideally, a database system should provide the support for high performance transaction processing, model complex applications, represent new kinds of data, and make intelligent deductions. While significant progress has been made during the late 1980s and early 1990s, there is much to be done before such a database system can be developed.

---

23 CAD/CAM stands for Computer Aided Design/Computer Aided Manufacturing.

24 For a discussion of the next generation database systems, we refer to [SIGM90].

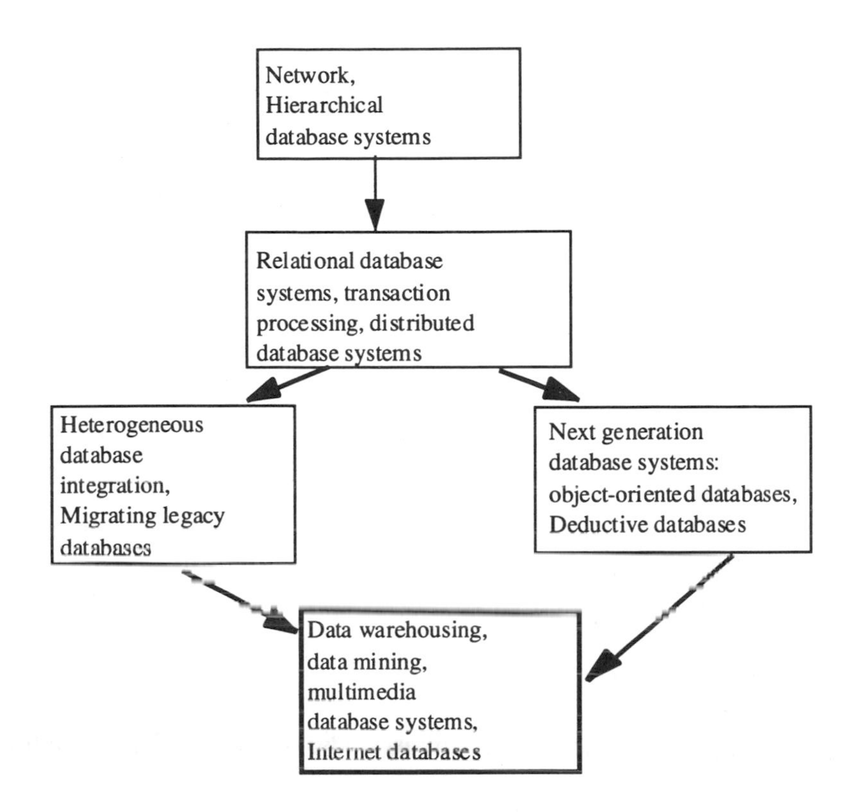

Figure A-1. Developments in Database Systems Technology

Heterogeneous database systems have been receiving considerable attention during the past decade [ACM90]. The major issues include handling different data models, different query processing strategies, different transaction processing algorithms, and different query languages. Should a uniform view be provided to the entire system or should the users of the individual systems maintain their own views of the entire system? These are questions that have yet to be answered satisfactorily. It is also envisaged that a complete solution to heterogeneous database management systems is a generation away. While research should be directed towards finding such a solution, work should also be carried out to handle limited forms of heterogeneity to satisfy the customer needs. Another type of database system that has received some attention lately is a federated database system. Note that some have used the terms heterogeneous database system and federated database system interchangeably. While heterogeneous database systems can be part of a federation, a federation can also include homogeneous database systems.

The explosion of users on the Internet as well as developments in interface technologies has resulted in even more challenges for data management researchers. A second workshop was sponsored by NSF in 1995, and several emerging technologies have been identified to be important as we go into the twenty-first century [NSF95]. These include digital libraries, managing very large databases, data administration issues, multimedia databases, data warehousing, data mining, data management for collaborative computing environments, and security and privacy. Another significant development in the 1990s is the development of object-relational systems. Such systems combine the advantages of both object-oriented database systems and relational database systems. Also, many corporations are now focusing on integrating their data management products with Internet technologies. Finally, for many organizations there is an increasing need to migrate some of the legacy databases and applications to newer architectures and systems such as client-server architectures and relational database systems. We believe there is no end to data management systems. As new technologies are developed, there are new opportunities for data management research and development.

A comprehensive view of all data management technologies is illustrated in Figure A-2. As shown, traditional technologies include database design, transaction processing, and benchmarking. Then there are database systems based on data models such as relational and object-oriented. Database systems may depend on features they provide such as security and real-time. These database systems may be relational or object-oriented. There are also database systems based on multiple sites or processors such as distributed and heterogeneous database systems, parallel systems, and systems being migrated. Finally, there are the emerging technologies such as data warehousing and mining, collaboration, and the Internet. Any comprehensive text on data management systems should address all of these technologies. We have selected some of the relevant technologies and put them in a framework. This framework is described in Section A.5.[25]

[25] In our previous book *Data Management Systems Evolution and Interoperation* we selected certain topics in data management and explained the various concepts.

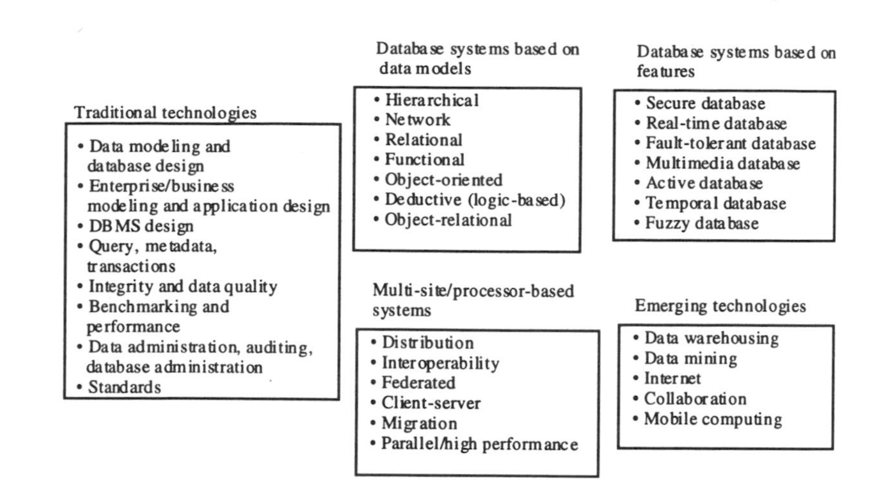

Figure A-2. Comprehensive View of Data Management Systems

A.3 STATUS, VISION AND ISSUES

Significant progress has been made on data management systems. However, many of the technologies are still stand-alone technologies as illustrated in Figure A-3. For example, multimedia systems are yet to be successfully integrated with warehousing and mining technologies. The ultimate goal is to integrate multiple technologies so that accurate data, as well as information, is produced at the right time and distributed to the user in a timely manner. Our vision for data and information management is illustrated in Figure A-4.

The work discussed in [THUR97] addressed many of the challenges necessary to accomplish this vision. In particular integration of heterogeneous databases, as well as the use of distributed object technology for interoperability, was discussed. While much progress has been made on the system aspects of interoperability, semantic issues still remain a challenge. Different databases have different representations. Furthermore, the same data entity may be interpreted differently at different sites. Addressing these semantic differences and extracting useful information from the heterogeneous and possibly multimedia data sources are major challenges. This book has attempted to address some of the challenges through the use of data mining.

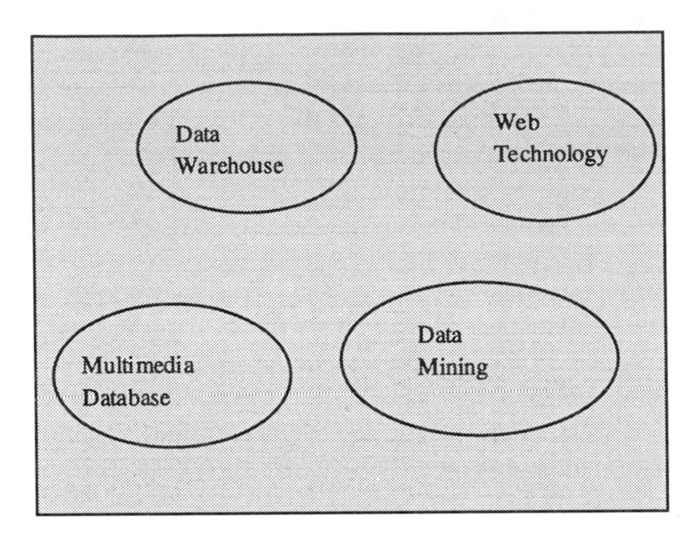

Figure A-3. Stand-alone Systems

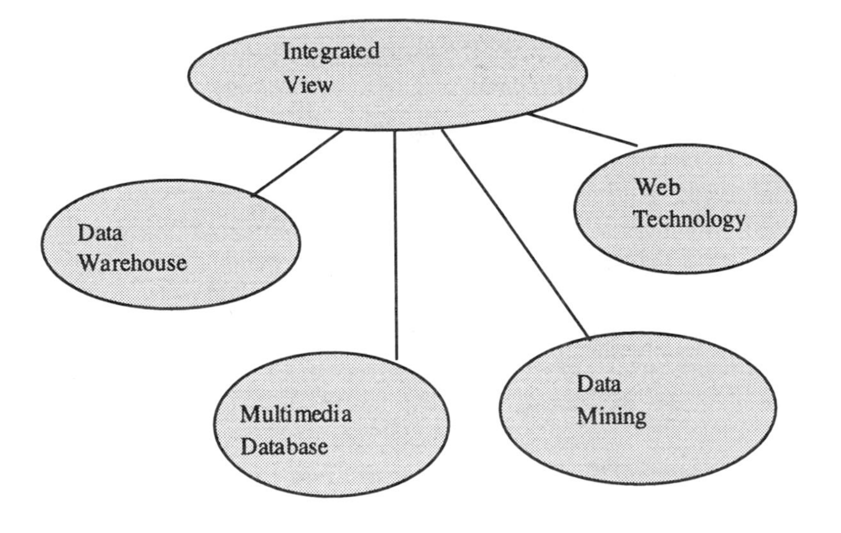

Figure A-4. Vision

A.4 DATA MANAGEMENT SYSTEMS FRAMEWORK

For the successful development of evolvable interoperable data management systems, heterogeneous database systems integration is a major component. However, there are other technologies that have to be successfully integrated with each other to develop techniques for efficient access and sharing of data as well as for the extraction of information from the data. To facilitate the development of data management systems to meet the requirements of various applications in fields such as medical, financial, manufacturing, and military, we have proposed a framework, which can be regarded as a reference model, for data management systems. Various components from this framework have to be integrated to develop data management systems to support the various applications.

Figure A-5 illustrates our framework, which can be regarded as a model, for data management systems.[26] This framework consists of three layers. One can think of the component technologies, which we will also refer to as components, belonging to a particular layer to be more or less built upon the technologies provided by the lower layer. Layer I is the Database Technology and Distribution Layer. This layer consists of database systems and distributed database systems technologies. Layer II is the Interoperability and Migration Layer. This layer consists of technologies such as heterogeneous database integration, client-server databases, multimedia database systems to handle heterogeneous data types, and migrating legacy databases.[27] Layer III is the Information Extraction and Sharing Layer. This layer essentially consists of technologies for some of the newer services supported by data management systems. These include data warehousing, data mining, Internet databases, and database support for collaborative applications.[28,29] Data management systems may utilize lower level technologies such as networking, distributed processing, and mass storage. We have grouped these technologies into a layer called the

---

26 Note that this three-layer model is subjective and is not a standard model. This model has helped us in organizing our views on data management.

27 We have placed multimedia database systems in Layer II, as we consider it to be a special type of a heterogeneous database system. A multimedia database system handles heterogeneous data types such as text, audio, and video.

28 Note that one could also argue whether database support for collaborative applications should be discussed here. This is because collaborative computing is not part of the data management framework. However, such applications do need database support, and our focus will be on this support.

29 Although Internet database management is an integration of various technologies, we have placed it in Layer III because it still deals with information extraction. Note that the data management framework consists of technologies for managing data as well as for extracting information from the data. However, what one does with the information, such as collaborative computing, sophisticated human computer interaction, natural language processing, and knowledge-based processing, does not belong to this framework. They belong to the Application Technologies Layer.

Supporting Technologies Layer. This supporting layer does not belong to the data management systems framework. This supporting layer also consists of some higher-level technologies such as distributed object management and agents.[30] Also, shown in Figure A-5 is the Application Technologies Layer. Systems such as collaborative computing systems and knowledge-based systems which belong to the Application Technologies Layer may utilize data management systems. Note that the Application Technologies Layer is also outside of the data management systems framework.

The technologies that constitute the data management systems framework can be regarded to be some of the core technologies in data management. However, features like security, integrity, real-time processing, fault tolerance, and high performance computing are needed for many applications utilizing data management technologies. Applications utilizing data management technologies may be medical, financial, or military, among others. We illustrate this in Figure A-6, where a three-dimensional view relating data management technologies with features and applications is given. For example, one could develop a secure distributed database management system for medical applications or a fault tolerant multimedia database management system for financial applications.[31]

Integrating the components belonging to the various layers is important to developing efficient data management systems. In addition, data management technologies have to be integrated with the application technologies to develop successful information systems. However, at present, there is limited integration between these various components. Our previous book *Data Management Systems Evolution and Interoperation* focused mainly on the concepts, developments, and trends belonging to each of the components shown in the framework. Furthermore, our current book on data mining focuses on the data-mining component of Layer 3 of the framework of Figure A-5.

---

30 Note that technologies such as distributed object management enable interoperation and migration.

31 In some cases one could also consider multimedia data processing and reengineering which is an essential part of system migration to be at the same level as features like security and integrity. One could also regard them to be emerging technologies.

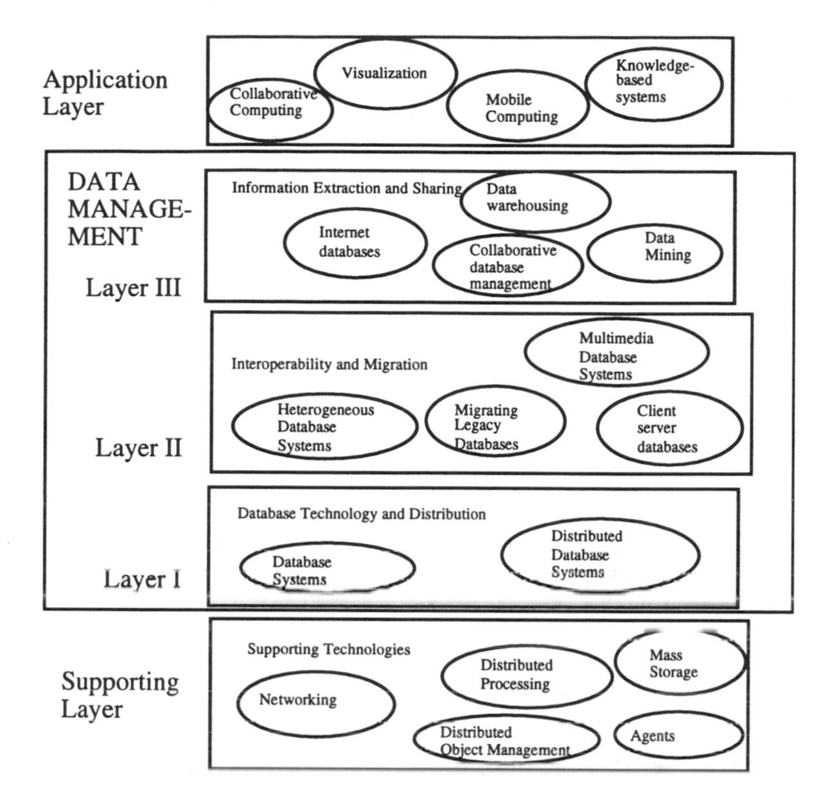

Figure A-5. Data Management Systems Framework

A.5 BUILDING INFORMATION SYSTEMS FROM THE FRAMEWORK

Figure A-5 illustrates a framework for data management systems. As shown in that figure, the technologies for data management include database systems, distributed database systems, heterogeneous database systems, migrating legacy databases, multimedia database systems, data warehousing, data mining, Internet databases, and database support for collaboration. Furthermore, data management systems take advantage of supporting technologies such as distributed processing and agents. Similarly, application technologies such as collaborative computing, visualization, expert systems, and mobile computing take advantage of data management systems.[32]

---

[32] Note that databases could also support expert systems as in the case of collaborative applications.

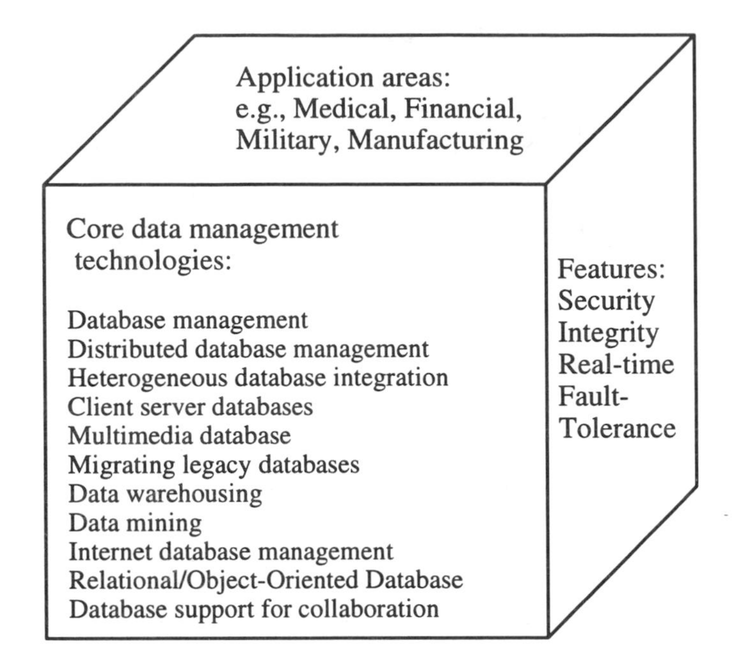

Figure A-6. A Three-dimensional View of Data Management

Many of us have heard of the term information systems on numerous occasions. These systems have sometimes been used interchangeably with data management systems. In our terminology, information systems are much broader than data management systems, but they do include data management systems. In fact, a framework for information systems will include not only the data management system layers, but also the supporting technologies layer as well as the application technologies layer. That is, information systems encompass all kinds of computing systems. It can be regarded as the finished product that can be used for various applications. That is, while hardware is at the lowest end of the spectrum, applications are at the highest end.

We can combine the technologies of Figure A-5 to put together information systems. For example, at the application technology level, one may need collaboration and visualization technologies so that analysts can collaboratively carry out some tasks. At the data management level, one may need both multimedia and distributed database technologies. At the supporting level, one may need mass storage as well as some distributed processing capability. This special framework is illustrated in Figure A-7. Another example is a special framework for interoperability. One may need some visualization technology to display the integrated information from the heterogeneous databases. At the data management level, we have heterogeneous database systems technology. At the supporting technology level, one may use distributed object management technology to encapsulate the heterogeneous databases. This special framework is illustrated in Figure A-8.

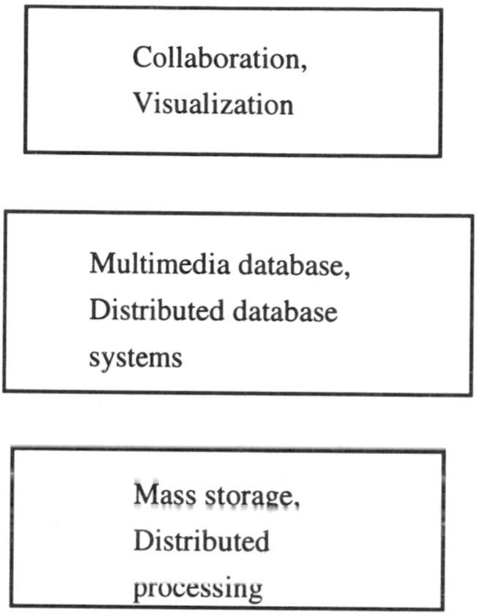

Collaboration,
Visualization

Multimedia database,
Distributed database
systems

Mass storage,
Distributed
processing

Figure A-7. Framework for Multimedia Data Management for Collaboration

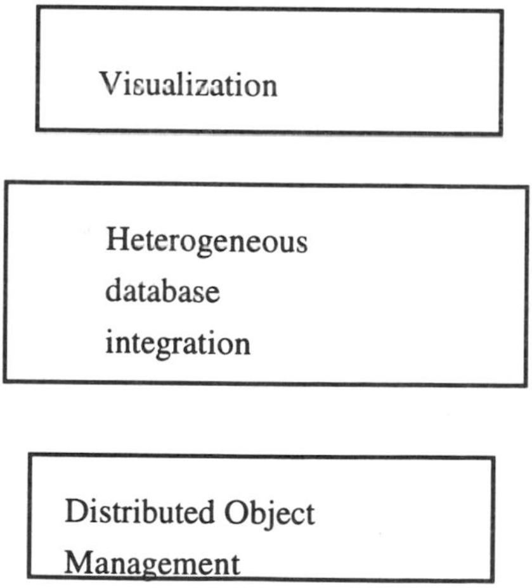

Visualization

Heterogeneous
database
integration

Distributed Object
Management

Figure A-8. Framework for Heterogeneous Database Interoperability

Finally, let us illustrate the concepts that we have described above by using a specific example. Suppose, a group of physicians/surgeons want a system where they can collaborate and make decisions about various patients. This could be a medical video teleconferencing application. That is, at the highest level, the application is a medical application and, more specifically, a medical video teleconferencing application. At the application technology level, one needs a variety of technologies including collaboration and teleconferencing. These application technologies will make use of data management technologies such as distributed database systems and multimedia database systems. That is, one may need to support multimedia data such as audio and video. The data management technologies in turn draw upon lower level technologies such as distributed processing and networking. We illustrate this in Figure A-9.

In summary, information systems include data management systems as well as application-layer systems such as collaborative computing systems and supporting-layer systems such as distributed object management systems.

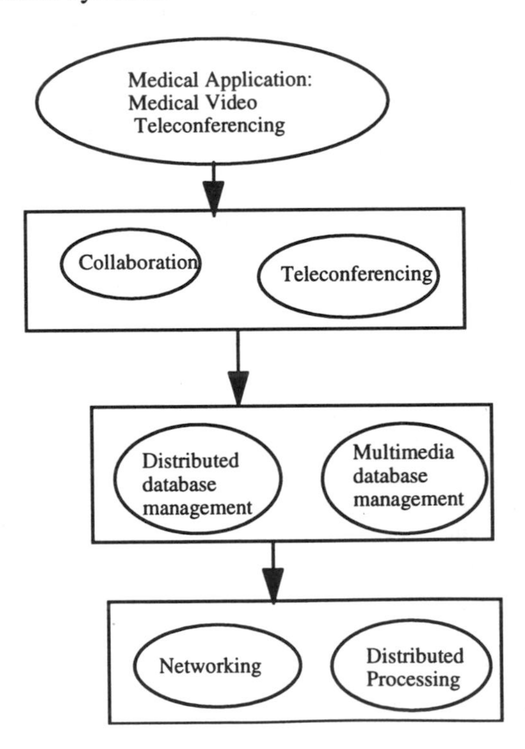

Figure A-9. Specific Example

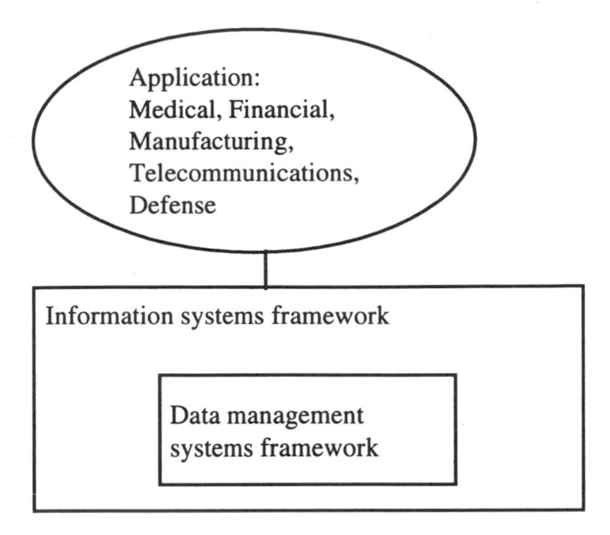

Figure A-10. Application-Framework Relationship

While application technologies make use of data management technologies and data management technologies make use of supporting technologies, the ultimate user of the information system is the application itself. Today numerous applications make use of information systems. These applications are from multiple domains such as medical, financial, manufacturing, telecommunications, and defense. Specific applications include signal processing, electronic commerce, patient monitoring, and situation assessment. Figure A-10 illustrates the relationship between the application and the information system.

A.6 SUMMARY

In this appendix we have provided an overview of data management. We first discussed the developments in data management and then provided a vision for data management. Then we illustrated a framework for data management. This framework consists of three layers: database systems layer, interoperability layer, and information extraction layer. Data mining belongs to Layer 3. Finally, we showed how information systems could be built from the technologies of the framework.

Let us repeat what we mentioned in Chapter 1 now that we have described the data management framework we introduced in [THUR97]. The chapters in this book not only discussed data mining concepts, they also showed how data mining could be applied to the various systems based on technologies of the framework we have shown in Figure A-5. For example, Chapter 2 described relationships between mining and databases. Chapter 3 described relationships between mining and warehousing. Chapter 10 described relationships between mining and distributed, heterogeneous, and legacy databases. Chapter 11 described relationships between mining and multimedia data. Chapter 12 described relationships between mining and the world wide web, which include digital libraries and Internet databases. That is, many of the technologies discussed in the framework of Figure A-5 have been useful in the discussion of mining. These include database systems, distributed database systems, interoperability of heterogeneous database systems, migrating legacy databases, multimedia database systems, data warehousing, and digital libraries and Internet database management. In addition, some other features for data management such as metadata, security, and logic programming also play a role in various chapters of this book. For example, metadata and mining was the subject of Chapter 14. Security and privacy issues were the subject of Chapter 13. Logic programming as a data mining technique was the subject of Chapter 8. Therefore, much of the discussions in this book have a strong orientation toward data and data management.

While data is the main concern for us, we have not ignored some of the other essential technologies and features of data mining. For example, Chapter 4 discussed other data mining technologies such as machine learning and statistics. Chapters 6 discussed steps to data mining and Chapter 7 described various data mining concepts and techniques. Chapter 9 provided an overview of the data mining tools. Therefore, we have tried to give a fairly balanced view of what is out there in data mining. Since artificial intelligence technology has also contributed much to data mining, we address this in Appendix B.

A.7 REFERENCES

[ACM90] Special Issue on Heterogeneous Database Systems, ACM Computing Surveys, September 1990.

[CATT91] Cattell, R., "Object Data Management Systems," Addison Wesley, MA, 1991.

[CODD70] Codd, E. F., "A Relational Model of Data for Large Shared Data Banks," Communications of the ACM, Vol. 13, #6, June 1970.

[DATE90] Date, C. J., "An Introduction to Database Management Systems," Addison Wesley, MA, 1990 (6th edition published in 1995 by Addison Wesley).

[MORE98] Morey, D., "Knowledge Management Architecture," Handbook of Data Management, Auerbach Publications, New York, 1998 (Ed: B. Thuraisingham).

[NSF90] Proceedings of the Database Systems Workshop, Report published by the National Science Foundation, 1990 (also in ACM SIGMOD Record, December 1990).

[NSF95] Proceedings of the Database Systems Workshop, Report published by the National Science Foundation, 1995 (also in ACM SIGMOD Record, March 1996).

[SIGM90] "Next Generation Database Systems," ACM SIGMOD Record," December 1990.

[THUR97] Thuraisingham, B., "Data Management Systems Evolution and Interoperation," CRC Press, FL, 1997.

APPENDIX B

ARTIFICIAL INTELLIGENCE

B.1 OVERVIEW

Much of the discussion in Appendix A has been on data management. As stated throughout this book we have taken a data-oriented perspective of data mining. As a result, we have given considerable attention to database systems, data warehousing and data management in general. However, data mining also has contributions from many other technologies. One such technology is machine learning. We provided an overview of machine learning in Chapter 4 and continued to discuss various machine learning techniques for data mining. Machine learning has roots in Artificial Intelligence, popularly known as AI. AI is all about making machines think and behave like humans. Researchers from various fields such as logic, philosophy, pattern recognition, language processing, speech processing, cognitive psychology and mathematics have contributed to artificial intelligence. It has roots in some of the early work in logic and philosophy starting in the nineteenth century and has subsequently involved researchers in other fields. AI has now evolved into various areas such as machine learning, knowledge management, expert systems, intelligent systems, and data mining and knowledge discovery. Perhaps the most significant paper in the history of AI is Alan Turing's paper on making machines think published in 1947.

Since AI technologies have played a major role in data mining, in this appendix we give an overview of AI. In Section B.2 we discuss the various developments in this field. We also discuss the integration of AI with other technologies such as database systems. The appendix is summarized in Section B.3. Finally some references are given in Section B.4.

B.2 DEVELOPMENTS IN ARTIFICIAL INTELLIGENCE TECHNOLOGIES

As stated in Section B.1, AI has roots in areas like logic, philosophy, cognitive psychology, and mathematics. It is all about making machines think and behave like human beings. Several excellent texts have been written on AI (see, for example, [WINS79]). Essentially, AI is an area that encompasses other areas like logic and philosophy. This section provides an overview of some of the developments.

In the nineteenth century, logicians started reasoning about formal systems. The development of first order predicate calculus was a significant milestone (see, for example, [MEND79]) where one was able to reason about formal systems. However, first order logic was found to be rigid and precise. In general, humans are much more flexible in their thinking. As a result, first order logic was extended to include non-standards logics such as modal logic and temporal logic. Various types of such logics have been applied to AI systems (see, for example, [TURN84]). While logicians started developing logics for AI, researchers such as Newell and Simon started developing ideas for machines to think. This was followed by the work of people like Feigenbaum and Minsky (see, for example, [MINS67]), This was when AI became an area on its own. Perhaps the most significant contribution to AI is Turing's famous paper on making machines think published in 1947. The early work in AI focused on developing programs to play games such as chess. These programs were in a way types of learning systems. They learned from previous experience, developed strategies, and learned to play games. One of the key developments here is intelligent searching. In order to effectively play games, it is impossible to search the entire space with all possibilities. Therefore, the challenge was on developing strategies that intelligently search, thereby limiting the search space.

While game playing programs were developed, a lot of work was carried out on developing semantic models. These models have also been used extensively in database management as discussed in Chapter 2. Examples of semantic models are frames and conceptual structures (see, for example, [SOWA84]). These models capture the real world entities and describe relationships between the entities.

Another area of development was in pattern matching and recognition, which was all about recognizing objects; natural language processing, where the system understands written language; and speech recognition, where the system understands spoken language. This whole area has now come to be known as human computer interaction which is all about building interfaces that improve communication between the human and the machine.

One of the significant developments of AI, mainly in the 1970s and beyond, is the development of expert systems. These are systems that perform the functions of experts. Subsequently, numerous expert systems have been developed for medical diagnosis, manufacturing, analyzing legal cases, and training students in various fields. Essentially expert systems perform the functions of people like physicianss, lawyers, teachers, and manufacturers.

Another area of development in AI is robotics. This topic deals with developing robots that perform the functions of various workers including surgeons, assembly line workers, people working in manufacturing plants, and pilots. With the emergence of the Internet, agent technology is now integrated heavily with robotics.

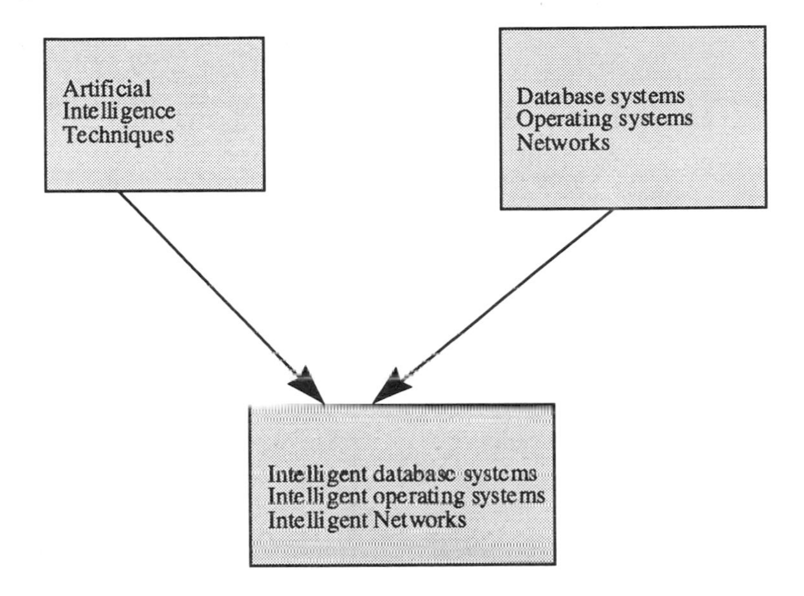

Figure B-1. Integration of Artificial Intelligence Techniques with Other Technologies

The developments in logic and AI led to the development of programming languages such as LISP and PROLOG (see, for example, [MCCA60] and [KOWA74]). LISP is a functional programming language and many expert systems in the early years were based on LISP. PROLOG is a logic programming language which uses logic for statements as discussed in Chapter 8. Logic programming evolved into inductive logic programming, which is now a data mining technique.

Finally, an area in AI that has contributed the most to data mining is machine learning. We gave an overview of machine learning in Chapter 4. As stated by Mitchell [MITC97], machine learning is about learning from experiences with respect to some performance measure. A good example of a machine learning program is a game playing program. Training in this case is the games that are used to train the system. The performance measure could be the number of games that are won by the system.

AI technologies have been integrated with a variety of technologies. The integration with database systems has produced intelligent database systems; the integration with logic has produced logic programming systems; and the integration with operating systems has produced intelligent operating systems. Essentially, AI has been integrated with several technologies to produce more intelligent systems. This integration is illustrated in figure B-1.

In summary, AI has evolved into a variety of areas such as human computer interaction, machine learning, training systems, and logic programming. Each of these areas has contributed to advancing the way machines perform and, as a result, have some value to data mining. The emergence of the Internet gives new opportunities for AI in getting the right information at the right time to the decision maker. The evolution of AI is illustrated in Figure B-2.

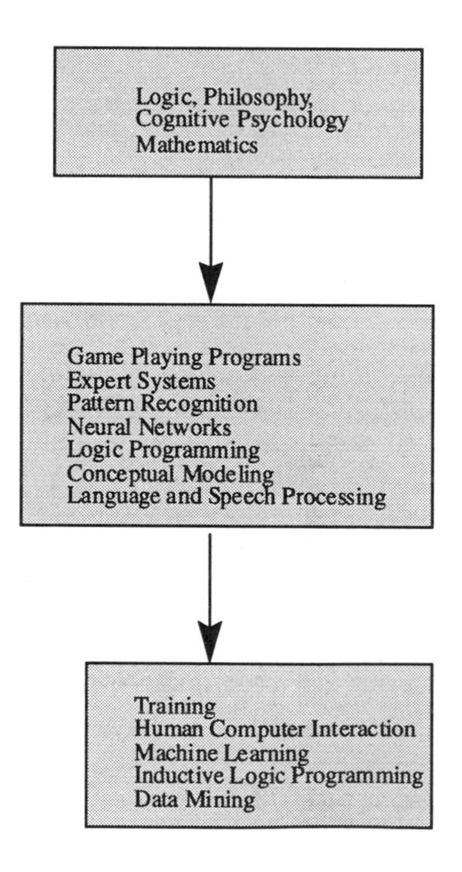

Figure B-2. Developments in Artificial Intelligence

B.3 SUMMARY

This appendix has provided a brief overview of the developments in AI. In particular, the origins of AI, the various developments such as game playing programs, expert systems, and robotics have been discussed. We also described the relationship of AI to machine learning, databases, logic programming, and data mining.

Although we do not hear as much about AI now as we did in the 1970s or 1980s, the developments that we have described here have had a significant impact on various technology areas. We now have various types of intelligent systems such as operating systems and database systems and even networks. Numerous expert systems have been developed. Furthermore, significant advances have been made in machine learning, training, and human computer interaction. They are all now contributing in some way to data mining.

B.4 REFERENCES

[KOWA74] Kowalski, R. A., "Predicate Logic as a Programming Language," Information Processing 74, Stockholm, North Holland Publications, 1974.

[MCCA60] McCarthy, J., "Recursive Functions of Symbolic Expressions and Their Computation by Machine," Communications of the ACM, Vol. 3, 1960..

[MEND79] Mendleson, E., "Mathematical Logic," Van Nostrand, NY, 1979.

[MINS67] Minsky, M., "Computation: Finite and Infinite Machines," Prentice Hall, NJ, 1967.

[MITC97] Mitchell, T., "Machine Learning," Addison Wesley, MA, 1997.

[SOWA84] Sowa, A., "Conceptual Structures: Information Processing in Minds and Machines," Addison Wesley. Reading, MA, 1984.

[TURN84] Turner, R., "Logics for Artificial Intelligence," Ellis Horwood, England, 1984.

[WINS79] Winston, P., "Artificial Intelligence," Addison Wesley, MA, 1979.

Index

Index